PRINCIPLES AND PRACTICE OF ENGINEERING

ENVIRONMENTAL ENGINEERING
SAMPLE QUESTIONS & SOLUTIONS

PRINCIPLES AND PRACTICE OF ENGINEERING
ENVIRONMENTAL ENGINEERING
SAMPLE QUESTIONS & SOLUTIONS

Published by the
National Council of Examiners for Engineering and Surveying®
280 Seneca Creek Road, Clemson, SC 29631 800-250-3196 www.ncees.org

ISBN-13: 978-1-932613-32-2
ISBN-10: 1-932613-32-3

Printed in the United States of America

TABLE OF CONTENTS

Introduction

One of the functions of the National Council of Examiners for Engineering and Surveying (NCEES) is to develop examinations that are taken by candidates for licensure as professional engineers. The NCEES has prepared this handbook to assist candidates who are preparing for the Principles and Practice of Engineering (PE) examination in environmental engineering. The NCEES is an organization established to assist and support the licensing boards that exist in all states and U.S. territories. The NCEES provides these boards with uniform examinations that are valid measures of minimum competency related to the practice of engineering.

To develop reliable and valid examinations, the NCEES employs procedures using the guidelines established in the *Standards for Educational and Psychological Testing* published by the American Psychological Association. These procedures are intended to maximize the fairness and quality of the examinations. To ensure that the procedures are followed, the NCEES uses experienced testing specialists possessing the necessary expertise to guide the development of examinations using current testing techniques.

The examinations are prepared by committees composed of professional engineers from throughout the nation. These engineers supply the content expertise that is essential in developing examinations. By using the expertise of engineers with different backgrounds such as private consulting, government, industry, and education, the NCEES prepares examinations that are valid measures of minimum competency.

Licensing Requirements

Eligibility

The primary purpose of licensure is to protect the public by evaluating the qualifications of candidates seeking licensure. While examinations offer one means of measuring the competency levels of candidates, most licensing boards also screen candidates based on education and experience requirements. Because these requirements vary among boards, it would be wise to contact the appropriate board. Board addresses and telephone numbers may be obtained by visiting our Web site at www.ncees.org or calling (800) 250-3196.

Application Procedures and Deadlines

Application procedures for the examination and instructional information are available from individual boards. Requirements and fees vary among the boards, and applicants are responsible for contacting their board office. Sufficient time must be allotted to complete the application process and assemble required data.

Description of Examinations

Examination Schedule

The NCEES PE examination in environmental engineering is offered to the boards in the spring and fall of each year. Dates of future administrations are as follows:

Year	Spring Dates	Fall Dates
2004	April 16	October 29
2005	April 15	October 28
2006	April 21	October 27

You should contact your board for specific locations of exam sites.

Examination Content

The 8-hour PE examination in environmental engineering is a no-choice examination consisting of 100 multiple-choice questions. The examination is administered in two 4-hour sessions each containing 50 questions. Each question has four answer options. The examination specifications presented in this book give details of the subjects covered on the examination.

Typically, all information required to answer a question is provided within the statement of the question itself. The question poses a scenario that an environmental engineer might encounter in practice soon after licensure. However, there may be instances where two to three questions are grouped within a single scenario with common information appearing before the questions. In these instances, the questions supply any further information specific to the question and define what is expected as a response to the question. In many cases, the correct response requires a calculation and/or conclusion that demonstrates competent engineering judgment.

A sample examination in environmental engineering is presented in this book. By illustrating the general content of the subject areas and formats, the questions should be helpful in preparing for the examination. Solutions are presented for all the questions. The solution presented may not be the only way to solve the question. The intent is to demonstrate the typical effort required to solve each question.

No representation is made or intended as to future examination questions, content, or subject matter.

EXAMINATION DEVELOPMENT

Examination Validity

Testing standards require that the questions on a licensing examination be representative of the important tasks needed for competent practice in the profession. The NCEES establishes the relationship between the examination questions and tasks by conducting an analysis of the profession that identifies the duties performed by the engineer. This information is used to develop an examination content outline that guides the development of job-related questions.

Examination Specifications

The examination content outline presented in this book specifies the subject areas that were identified for environmental engineering and the percentage of questions devoted to each of them. The percentage of questions assigned to each of the subject areas reflects both the frequency and importance experienced in the practice of environmental engineering.

Examination Preparation And Review

Examination development and review workshops are conducted at least twice annually by standing committees of the NCEES. Additionally, workshops are held as required to supplement the bank of questions available. The content and format of the questions are reviewed by the committee members for compliance with the specifications and to ensure the quality and fairness of the examination. These engineers are selected with the objective that they be representative of the profession in terms of geography, ethnic background, gender, and area of practice.

Minimum Competency
One of the most critical considerations in developing and administering examinations is the establishment of passing scores that reflect a standard of minimum competency. The concept of minimum competency is uppermost in the minds of the committee members as they assemble questions for the examination. Minimum competency, as measured by the examination component of the licensing process, is defined as the lowest level of knowledge at which a person can practice professional engineering in such a manner that will safeguard life, health, and property and promote the public welfare.

To accomplish the setting of fair passing scores that reflect the standard of minimum competency, the NCEES conducts passing score studies on a periodic basis. At these studies, a representative panel of engineers familiar with the candidate population uses a criterion-referenced procedure to set the passing score for the examination. Such procedures are widely recognized and accepted for occupational licensing purposes. The panel discusses the concept of minimum competence and develops a written standard of minimum competency that clearly articulates what skills and knowledge are required of licensed engineers. Following this, the panelists take the examination and then evaluate the difficulty level of each question in the context of the standard of minimum competency.

The NCEES does not use a fixed-percentage pass rate such as 70% or 75% because licensure is designed to ensure that practitioners possess enough knowledge to perform professional activities in a manner that protects the public welfare. The key issue is whether an individual candidate is competent to practice and not whether the candidate is better or worse than other candidates.

The passing score can vary from one administration of the examination to another to reflect differences in difficulty levels of the examinations. However, the passing score is always based on the standard of minimum competency. To avoid confusion that might arise from fluctuations in the passing score, scores are converted to a standard scale that adopts 70 as the passing score. This technique of converting to a standard scale is commonly employed by testing specialists.

Scoring Procedures

The examination consists of 100 equally weighted multiple-choice questions. There is no penalty for marking incorrect responses; therefore candidates should answer each question on the examination. Only one response should be marked for each question. No credit is given where two or more responses are marked. The examination is compensatory—poor scores in some subjects can be offset by superior performance elsewhere.

The legal authority for making licensure decisions rests with the individual licensing boards and not with the NCEES. Consequently, each board has the authority to determine the passing score for the examination. The NCEES provides each board with a recommended passing score based on the criterion-referenced procedure described previously.

Examination Procedures and Instructions

Examination Materials
Before the morning and afternoon sessions, proctors will distribute examination booklets containing an answer sheet. You should not open the examination booklet until you are instructed to do so by the proctor. Read the instructions and information given on the front and back covers, shown in

Appendix A. Enter your name in the upper right corner of the front cover. Listen carefully to all the instructions the proctor reads.

The answer sheets for the multiple-choice questions are machine scored. For proper scoring, the answer spaces should be blackened completely. Since April 2002, the NCEES has provided mechanical pencils with 0.7-mm HB lead to be used in the examination. You are not permitted to use any other writing instrument. If you decide to change an answer, you must erase the first answer completely. Incomplete erasures and stray marks may be read as intended answers. One side of the answer sheet is used to collect identification and biographical data. Proctors will guide you through the process of completing this portion of the answer sheet prior to taking the test. This process will take approximately 15 minutes.

Starting and Completing the Examination

You are not to open the examination booklet until instructed to do so by your proctor. If you complete the examination with more than 15 minutes remaining, you are free to leave after returning all examination materials to the proctor. Within 15 minutes of the end of the examination, you are required to remain until the end to avoid disruption to those still working and to permit orderly collection of all examination materials. Regardless of when you complete the examination, you are responsible for returning the numbered examination booklet assigned to you. Cooperate with the proctors collecting the examination materials. Nobody will be allowed to leave until the proctors have verified that all materials have been collected.

References

The PE examination is open-book. Your board determines the reference materials and calculators that will be allowed. In general, you may use textbooks, handbooks, bound reference materials, and a non-communicating, battery-operated, silent, non-printing calculator. States differ in their rules regarding calculators and references, and you should contact your board for specific advice. The NCEES Web site at www.ncees.org also provides guidelines and information about calculator use on examinations.

Special Accommodations

If you require special accommodations in the test-taking procedure, you should communicate your need to your board office well in advance of the day of the examination so that necessary arrangements may be made.

UPDATES TO EXAMINATION INFORMATION

For updated exam specifications, errata for this book, and other information about exams, visit the NCEES Web site at www.ncees.org.

EXAMINATION SPECIFICATIONS

THE NATIONAL COUNCIL OF EXAMINERS FOR ENGINEERING AND SURVEYING

PRINCIPLES AND PRACTICE OF ENGINEERING EXAMINATION

ENVIRONMENTAL

EFFECTIVE October 2004

	Approximate Percentage of Examination	New 2011 test
I. Water	**34%**	32%
A. Wastewater	11%	10

1. Sources of pollution and minimization/prevention
2. Treatment technologies
3. Water chemistry
4. Sampling and measurement methods
5. Biology/microbiology
6. Hydraulics/fluid mechanics
7. Fate and transport
8. Collection systems
9. Residuals management (solid, liquid, and gas)
10. Codes, standards, regulations, and guidelines
11. Mathematics and statistics
12. Engineering economics

B. Storm Water	6%	4

1. Sources of pollution
2. Hydrology/hydrogeology
3. Hydraulics/fluid mechanics
4. Sampling and measurement methods
5. Treatment technologies
6. Collection systems
7. Fate and transport
8. Codes, standards, regulations, and guidelines
9. Mathematics and statistics
10. Engineering economics

C. Potable Water	11%	10

1. Treatment technologies
2. Water chemistry
3. Hydraulics/fluid mechanics
4. Biology/microbiology
5. Sampling and measurement methods

Approximate Percentage of Examination

6. Distribution systems
7. Residuals management
8. Codes, standards, regulations, and guidelines
9. Mathematics and statistics
10. Engineering economics

D. Water Resources — 6%

1. Sources of pollution
2. Hydrology/hydrogeology
3. Water chemistry
4. Biology/microbiology
5. Sampling and measurement methods
6. Hydraulics and limnology
7. Fate and transport
8. Codes, standards, regulations, and guidelines
9. Mathematics and statistics
10. Engineering economics
11. Watershed management and planning

II. Air — 20%

A. Ambient Air — 8%

1. Meteorology
2. Atmospheric chemistry
3. Sampling and measurement methods
4. Risk assessment
5. Codes, standards, regulations, and guidelines
6. Mathematics and statistics

B. Emissions Sources — 4%

1. Chemistry
2. Aerosol science
3. Thermodynamics
4. Sampling and measurement methods
5. Fate and transport (dispersion)
6. Source categories
7. Odor generation and control
8. Sources of pollution and minimization/prevention
9. Codes, standards, regulations, and guidelines
10. Mathematics and statistics
11. Engineering economics

	Approximate Percentage of Examination
C. Control Strategies	8%

1. Treatment technologies
2. Air transport systems
3. Thermodynamics
4. Aerosol science
5. Chemistry
6. Sampling and measurement methods
7. Sources of pollution and minimization/prevention
8. Codes, standards, regulations, and guidelines
9. Mathematics and statistics
10. Engineering economics

III. Solid, Hazardous, and Special Waste — **20%**

A. Municipal Solid Waste (MSW), Commercial, and Industrial Wastes — 10%

1. Definition and characterization of different types of solid waste
2. Sampling and measurement methods
3. Storage, collection, and transportation systems
4. Minimization, reduction, and recycling
5. Risk assessment
6. Fate and transport
7. Treatment and disposal technologies
8. Chemistry
9. Codes, standards, regulations, and guidelines
10. Engineering economics

B. Hazardous Waste, Special, and Radioactive Waste — 10%

1. Definition and characterization of different types of waste
2. Sampling and measurement methods
3. Storage, collection, and transportation systems
4. Minimization, reduction, and recycling
5. Risk assessment
6. Fate and transport
7. Treatment and disposal technologies
8. Chemistry
9. Health physics
10. Codes, standards, regulations, and guidelines
11. Mathematics and statistics
12. Engineering economics

Approximate Percentage of Examination

IV. Environmental Assessments, Remediation, and Emergency Response — **26%**

A. Environmental Assessments — 8%

1. Site assessment
2. Hydrogeology
3. Sampling and measurement methods
4. Historical considerations and land use practices
5. Fate and transport
6. Sources of pollution
7. Exposure/risk characterization
8. Codes, standards, regulations, and guidelines

B. Remediation — 8%

1. Remediation alternatives
2. Minimization/prevention
3. Hydrology/hydrogeology
4. Codes, standards, regulations, and guidelines
5. Engineering economics
6. Sampling and measurement methods
7. Control technologies

C. Public Health and Safety — 10%

1. Industrial hygiene, health, and safety
2. Security, emergency plans and incident response procedures
3. Fundamentals of epidemiology and toxicology
4. Exposure assessments
5. Radiation protection/health physics
6. Vector control and sanitation including biohazards
7. Noise pollution
8. Indoor air quality
9. Codes, standards, regulations, and guidelines

TOTAL 100%

NOTES:

1. The knowledge areas specified under A, B, C, etc., are examples of kinds of knowledge, but they are not exclusive or exhaustive categories.
2. This examination contains 100 multiple-choice questions. Examinee works all questions.

SAMPLE QUESTIONS FOR THE MORNING PORTION OF THE EXAMINATION IN ENVIRONMENTAL ENGINEERING

Subject	Questions
Water	
Wastewater	11
Storm Water	6
Potable Water	11
Water Resources	6
Air	
Ambient Air	8
Emissions Sources	4
Control Strategies	4
Total:	50

101. An existing single-stage rock-media trickling filter will be upgraded by addition of a rotating biological contactor (RBC) process. The flow diagram is shown in the figure below. Trickling filter performance is described by the National Research Council (NRC) equation as shown below.

$$E = \frac{100}{1 + 0.0561\sqrt{W/VF}}$$

E = Efficiency of BOD_5 removal across filter including clarifier (%)

W = BOD_5 influent loading to filter (lb/day)

V = Volume of filter (1,000 ft^3)

F = Recirculation factor

$$= \frac{1 + R}{(1 + R/10)^2}$$

R = Recirculation ratio = Q_R/Q

The influent has a biological oxygen demand (BOD_5) concentration of 250 mg/L. An effluent BOD_5 of 15 mg/L is required from the treatment plant.

The influent flow rate is 1.5 MGD, and the recirculation ratio is 0.5. The equation for BOD_5 removal through the RBC process (including the final clarifier) is:

$$\frac{S_e}{S_o} = \frac{1}{[1 + KA/Q]^3}$$

where:
S_e = BOD_5 following the RBC process, mg/L
S_o = BOD_5 applied to RBC process, mg/L
K = constant = 2.45 gpd/ft^2
A = RBC disc area for single stage, ft^2
Q = Hydraulic flow rate, gpd

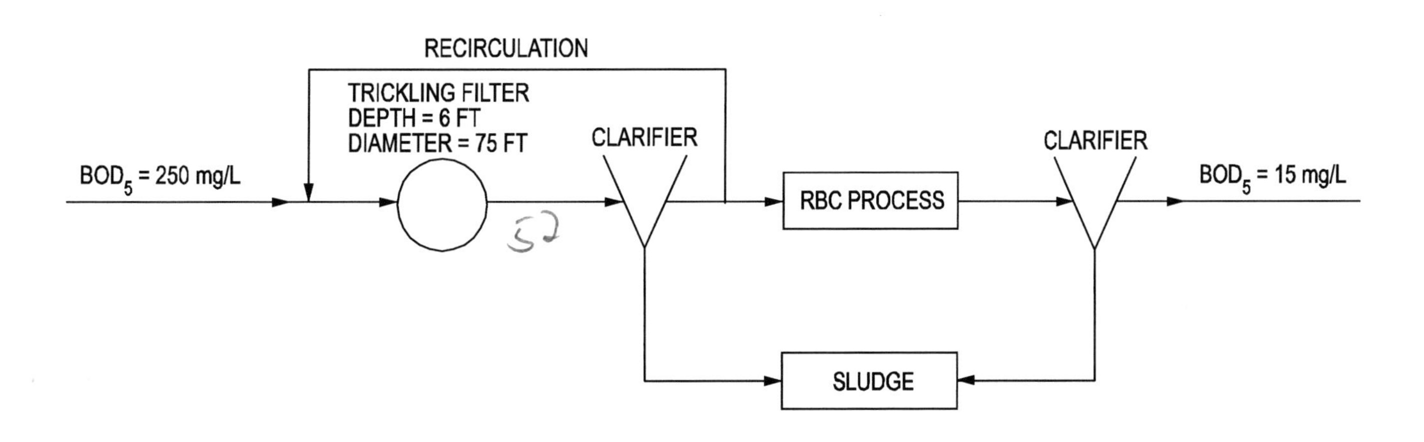

101. (Continued)

The BOD_5 loading (lb/day) to the RBC unit is most nearly:

(A) 1,075
(B) 1,300
(C) 1,600
(D) 2,000

102. A conventional activated sludge plant, as shown in the figure below, treats 2.0 MGD. The raw wastewater contains 250 mg/L suspended solids and 200 mg/L BOD_5. The aeration tank volume is 1,500,000 gallons, and mixed liquor suspended solids (MLSS) is 2,800 mg/L. The recycled sludge flow is 800,000 gpd containing 8,000 mg/L suspended solids. The suspended solids loading to the secondary clarifier is 8.0 lb/ft^2/day, and the surface loading to the primary clarifier is 720 gallons/ft^2/day.

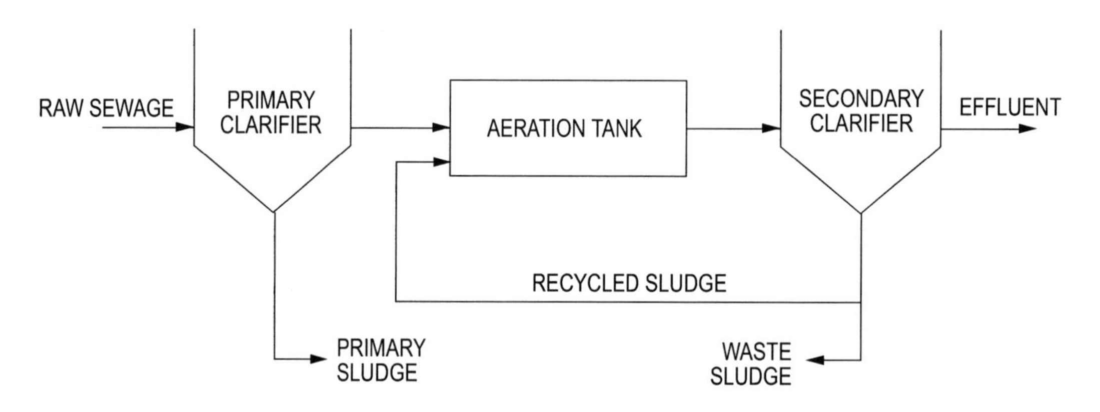

ACTIVATED SLUDGE PLANT – SCHEMATIC

If the BOD_5 removal in the primary clarifier is 40% and in the aeration tank is 92%, the efficiency of the BOD_5 removal in the treatment plant is most nearly:

(A) 37%
(B) 77%
(C) 92%
(D) 95%

103. The presence of heavy metals in dewatered wastewater sludge would cause the following concern:

(A) poor dewatering characteristics leading to wet sludge
(B) chemical reactions between the heavy metals in the soil
(C) binding excess water in the soil during winter months
(D) elimination of land application as a viable disposal alternative

104. According to EPA protocol, during a wastewater 24-hour composite sampling event, which of the following must be taken as a grab sample?

(A) BOD
(B) TSS
(C) Metals
(D) Oil and grease

105. During an analysis of a wastewater sample for TKN, the ammonia concentration was found to be zero. What could be wrong?

(A) The pH was not elevated prior to distillation.
(B) The pH was not decreased prior to distillation.
(C) The catalyst for the digestion sample was forgotten.
(D) The acid digestion period was carried past the recommended period.

106. Assume a 12-in. concrete ($n = 0.013$) trunk sewer has a slope of 0.004. The minimum depth (in.) required to maintain a self-cleansing velocity of 2 fps is:

(A) 4.5
(B) 7.0
(C) 8.4
(D) impossible to achieve a cleansing velocity at this slope

107. A 4.0-MGD wastewater treatment plant discharges a secondary-treated effluent into a receiving stream.

The wastewater has the following characteristics:

BOD_5	20 mg/L
Dissolved oxygen	2.0 mg/L
Water temperature	24°C

The receiving stream upstream from the point of wastewater discharge has the following characteristics:

Flow rate	18 cfs
BOD_5	4.0 mg/L
Dissolved oxygen	6.0 mg/L
Water temperature	27°C
Cross-sectional area	36 ft^2 (uniform)

The reaeration rate is estimated to be 0.4 day^{-1} (base e at 20°C), and the temperature correction coefficient is 1.024.

The deoxygenation rate constant is estimated to be 0.23 day^{-1} (base e at 20°C), and the temperature correction coefficient is 1.047.

Reaeration and deoxygenation are the only major factors affecting the dissolved oxygen concentration in the stream after mixing with wastewater effluent.

Assume the time of travel in the stream to reach the maximum dissolved oxygen deficit is 3 days. The dissolved oxygen deficit (mg/L) at this location downstream from the point of discharge is most nearly:

(A) 3.2
(B) 5.2
(C) 7.2
(D) 8.4

108. The following information applies:

Acute Criterion
$X\ (\mu g/L) = \exp[M_A\ (\ln(\text{Hardness})) + b_A]$

Chronic Criterion
$X\ (\mu g/L) = \exp[M_C\ (\ln(\text{Hardness})) + b_C]$

Hardness-Dependent Aquatic Toxicity Criteria				
Metal	Acute Criterion		Chronic Criterion	
	M_A	b_A	M_C	b_C
Cd	1.128	–3.828	0.785	–3.490

K_{sp} of $Cd(OH)_2 = 5.33 \times 10^{-15}$; MW of cadmium = 112.4 gm/mol

Downstream of the metal plating facility lagoon holding wastewater and sludges, surface water monitoring data indicate 4-day and 1-hour average cadmium concentrations of 6.8 µg/L and 14.2 µg/L, respectively. An average water hardness value of 125 mg/L as $CaCO_3$ was also measured. Given the data in the table the cadmium criterion (µg/L) for long-term ecological protection of aquatic life is most nearly:

(A) 1.4
(B) 5.0
(C) 14.5
(D) 22.0

109. A laboratory reports the following from a sludge feed sample:

Weight of dish	19.15 g
Weight of dish and sample	133.61 g
Weight of dish and dry solids	25.72 g
Weight of dish and ash	20.37 g

The % solids and % volatile solids, respectively, are most nearly:

(A) 19.3%, 81.4%
(B) 5.7%, 81.4%
(C) 5.7%, 18.6%
(D) 5.7%, 4.7%

110. A constant volume batch reactor was operated to provide data for determination of reaction rate constant for ultraviolet radiation enhanced ozonation of a synthetic organic compound. The duration of the batch reaction was 1 hr and the concentration of the organic compound was reduced from 200 μg/L to 10 μg/L.

If the reaction is assumed to be first-order with respect to the synthetic organic compound, then the rate constant for the oxidation reaction is:

(A) 3.0 hr^{-1}
(B) 3.0 μg/(L-hr)
(C) 190 hr^{-1}
(D) 190 μg/(L-hr)

111. Chlorination/ dechlorination and ultraviolet light disinfection systems are being considered for a new municipal wastewater treatment plant . The NPDES permit requires chlorine to be ≤ 0.019 mg/L and *E. coli* ≤ 200 MPN/100 mL. Capital and operating costs are as follows:

Item	Chlor/Dechlor	UV
Capital	$600,000	$750,000
Labor	30 hr/month	50 hr/month
Power	1,500 W/hr	300 75-W lamps
Hypochlorite usage	100 gpd	
Bisulfite usage	10 gpd	
Lamp replacement		50%/yr
Misc. OM & R	$3,000/yr	$3,000/yr

Unit costs are:

Labor	$20/hr
Power	$0.07/kWh
Hypochlorite	$0.87/gal
Bisulfite	$192.50/55-gal drum
UV lamps	$40 each

Present worth discount rate is 6%.

111. (Continued)

Perform a cost evaluation of the two systems. Due to planned regionalization, the proposed system will be replaced in 10 yr. Which of the following represents the resulting process selection:

(A) The chlorination/dechlorination system is more cost effective and should be selected for design.
(B) The ultraviolet light system is more cost effective and should be selected for design.
(C) Within the accuracy of the analysis the two systems have equal cost. The design decision must be made on non-cost basis.
(D) The chlorination system should be selected since, if UV is selected, it will still be necessary to feed chlorine to provide a residual in the distribution system.

112. A well-maintained, straight roadside drainage ditch (shown in the figure below) with short vegetation cover ($n = 0.030$) is to carry a maximum flow of 3 cfs over a reach of 1,200 feet and a gradual vertical drop of 1.2 feet. The triangular ditch is to be constructed with side slopes of 4:1 (H:V) in sandy loam soil.

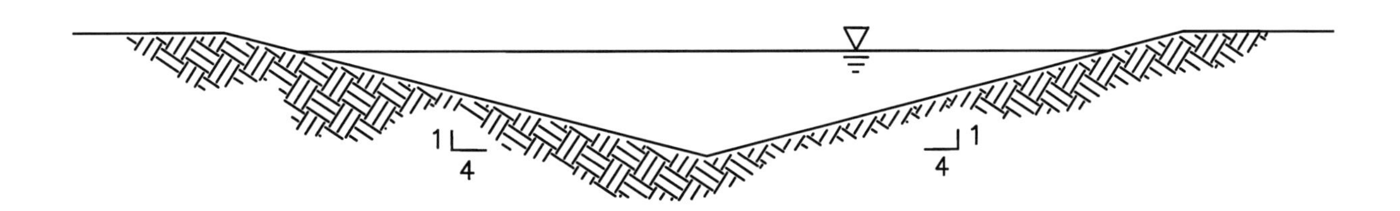

DRAINAGE DITCH SECTION
NOT TO SCALE

If the flow is 2 cfs, the critical depth (feet) is most nearly:

(A) 0.21
(B) 0.38
(C) 0.43
(D) 0.57

Questions 113–114: Figure 1 depicts a drainage basin. **Figure 2** and the table on the opposite page provide hydrologic parameters for the basin.

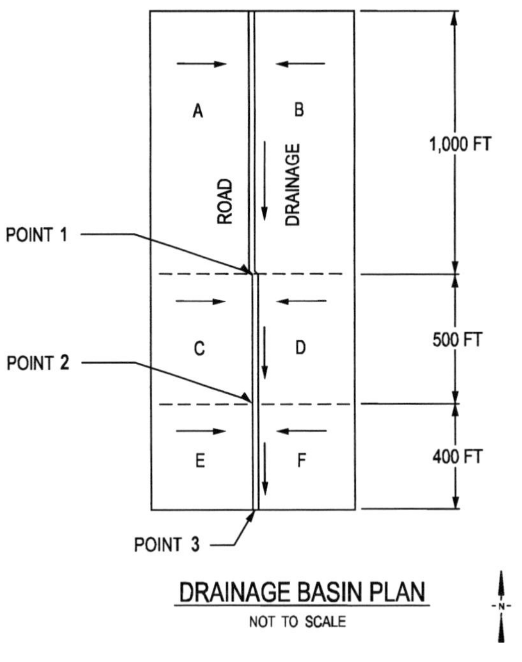

FIGURE 1

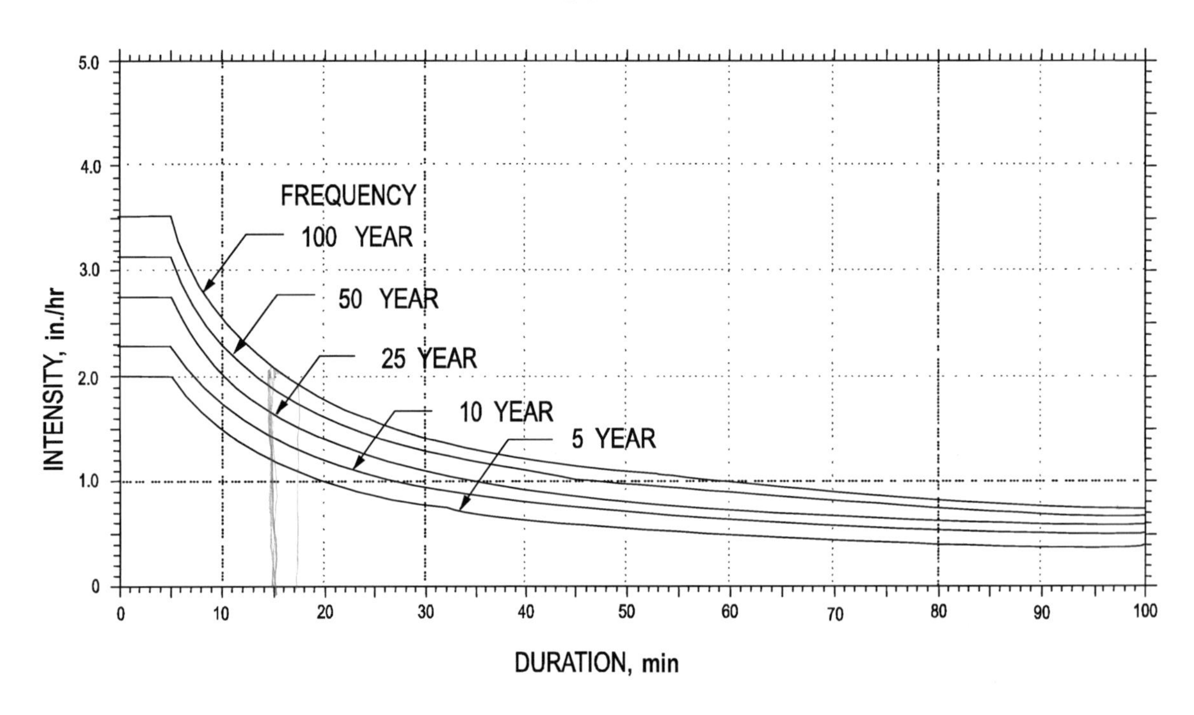

RAINFALL INTENSITY, DURATION, FREQUENCY

FIGURE 2

GO ON TO THE NEXT PAGE

Questions 113–114 (Continued)

Area	Land Use	Ave Number of Residences/Acre	Area (acres)	NRCS/SCS Runoff Curve Number
A	Forest (Good Cover)	0	8	55
B	Forest (Good Cover)	0	12	55
C	Park/Open Space (Good Condition)	0	6	61
D	Residential	2	6	75
E	Residential	2	4	75
F	Residential	4	4	85

113. Assume all soils in the drainage basin are in hydrologic soil group B according to the Natural Resource Conservation Service (NRCS), formerly known as the Soil Conservation Service (SCS). Also assume that the vegetative covers in areas identified as A, B, and C are in good condition. The NRCS/SCS Runoff Curve Number (CN) for the entire area is most nearly:

(A) 60
(B) 65
(C) 70
(D) 75

114. The following table gives overland flow times for different parts of the drainage basin:

Segment of Drainage Basin	Overland Flow Time (min)
Area A to Point 1	5
Area B to Point 1	6
Area C to Point 2	10
Area D to Point 2	12
Area E to Point 3	3
Area F to Point 3	4
Point 1 to Point 2	2
Point 2 to Point 3	3

For a 100-year design storm, the design rainfall intensity (in./hr) to be used at Point 3 is most nearly:

(A) 2.0
(B) 2.5
(C) 3.0
(D) 3.5

115. The velocity of the water in a concrete channel is a function of which of the following?

I. Roughness coefficient
II. Channel slope
III. Cross-sectional area

(A) I and II only
(B) I and III only
(C) II and III only
(D) I, II, and III

116. The following flows were measured at the point of discharge of a drainage basin.

Time (hr)	Storm Water Discharge (cfs)
0	0
2	2.5
4	5
6	7.5
8	4.0
10	2.0
12	1.0
14	0

For a 2-hour duration storm and a drainage area of 40 acres, the unit hydrograph discharge (cfs) after 10 hours would be most nearly:

(A) 1.1
(B) 1.8
(C) 2.5
(D) 3.0

117. Stormwater discharges are now to be regulated by EPA. Although first charged with the task in 1972, it was the 1987 amendments that initiated action. Initially, industries could file group applications. What information was required for a group application, Part I?

(A) topographical maps and locations of chemicals
(B) identification of refuse site
(C) information concerning quantities of waste produced
(D) information concerning participants and their industrial activities

118. Disinfection is to be added to a wastewater treatment plant with an average daily flow of 13 MGD and an hourly peaking factor of 2.5. The average coliform count (N_o), must be reduced from 10,000 org/100 ml to 200 org/100 ml (N_t). The temperature of the wastewater is 60°F.

The planned facilities shown in the figure below include a gas chlorinator, a horizontal channel for a hydraulic jump for mixing, and a detention basin.

Codes require a 15-minute contact time for peak hourly flow, a 30-minute contact time for average daily flow, and a maximum chlorine dosage capability of 15 mg/L.

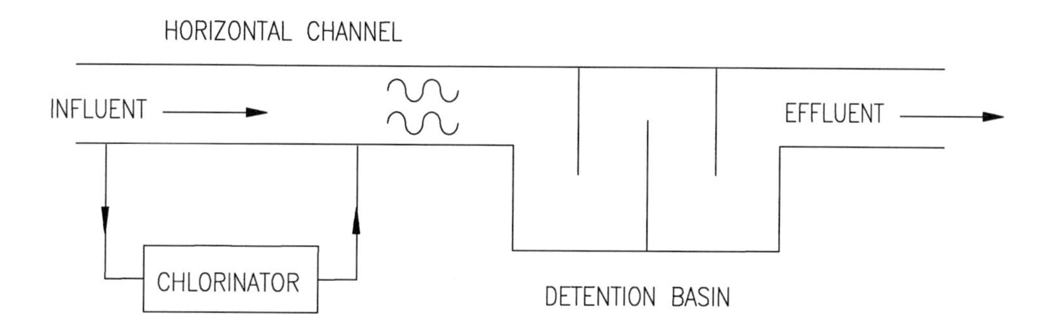

If the chlorine demand is 11.0 mg/L (as Cl_2) and the required free available total chlorine residual is 0.5 mg/L, then the smallest commercially available chlorinator (lb/hr) that will provide the required total dosage is:

(A) 50
(B) 75
(C) 100
(D) 150

GO ON TO THE NEXT PAGE

119. The results of a well water analysis are given below.

Ca^{+2}	51 mg/L
Mg^{+2}	12 mg/L
Na^{+}	25 mg/L
SO_4^{-2}	65 mg/L
Cl^{-}	25 mg/L
F^{-}	0.4 mg/L
NO_3^{-}	14 mg/L as N
pH	7.8
H_2S	3.4 mg/L as S
Alkalinity	84 mg/L as $CaCO_3$
Total coliforms	MPN 2.2 org/100 ml
Turbidity	6.2 NTU
Chlorine demand	9.1 mg/L
TDS	332 mg/L
Temperature	25°C

The difference (meq/L) between the sum of the cations and sum of the anions is most nearly:

(A) 0.1
(B) 0.3
(C) 0.5
(D) 0.7

GO ON TO THE NEXT PAGE

120. Water is pumped between two reservoirs as shown in **Figure 1**. The pump characteristics are given in **Figure 2**. The Hazen-Williams formula ($V=1.318\ CR^{0.63}\ S^{0.54}$) is to be used to estimate friction losses. The Hazen-Williams coefficient is 100.

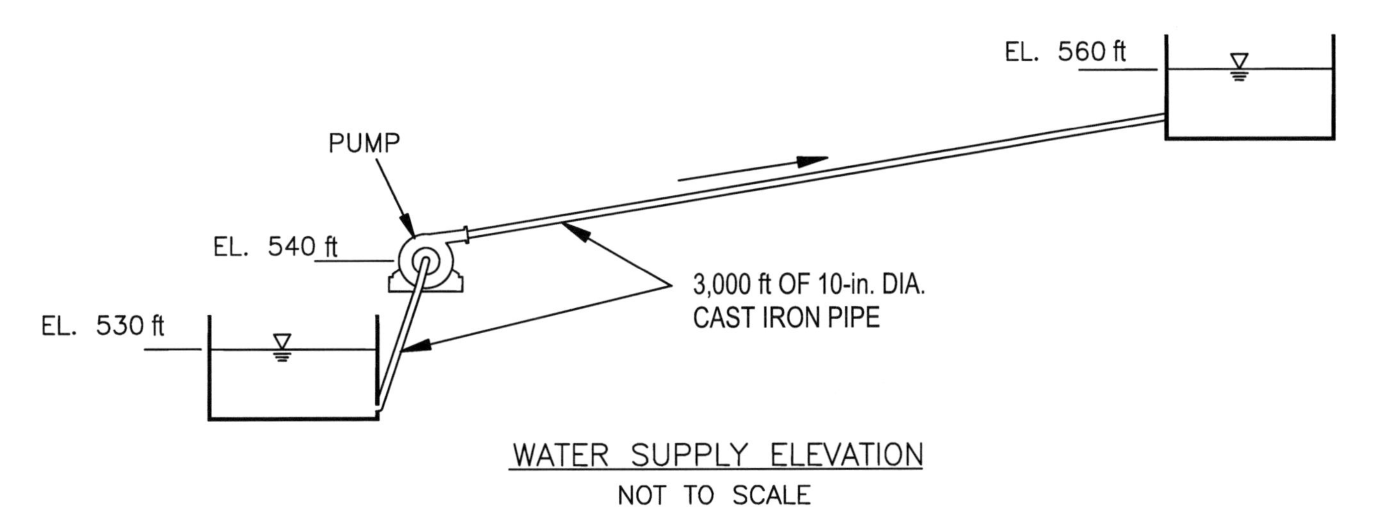

FIGURE 1

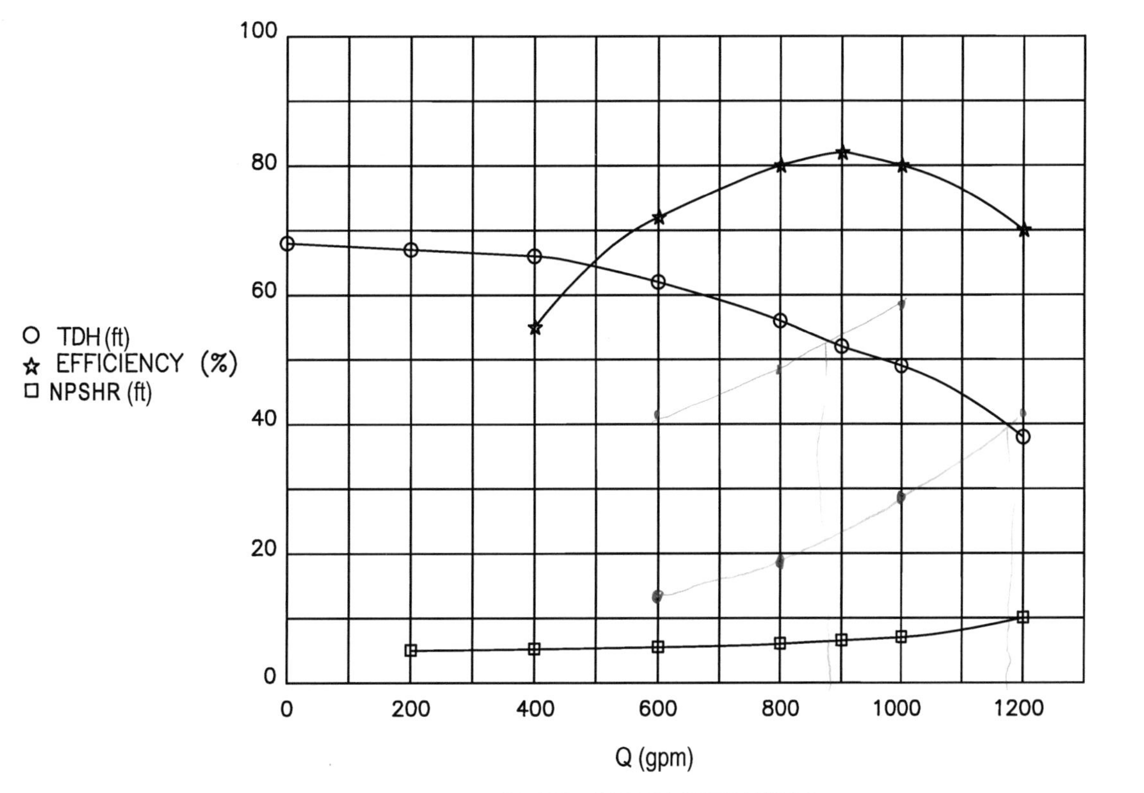

FIGURE 2

120. (Continued)

The flow (gpm) through the existing system will be most nearly:

(A) 510
(B) 650
(C) 870
(D) 1,030

121. Which of the following will reduce the tendency for pump cavitation?

I. Increasing the discharge pipe diameter
II. Lowering the pump elevation
III. Increasing the suction pipe diameter

(A) II and III only
(B) II only
(C) I and II only
(D) I, II, and III

122. A 1-ft-diameter well is constructed into the confined aquifer shown in the figure below. The aquifer has a uniform thickness at 30 ft overlain by an impermeable layer 50 ft deep. The initial piezometric surface is 38 ft below the ground surface datum of the test well and observation wells. After water was pumped for several days at 0.3 cfs, the water levels stabilized at the following drawdowns:

Test well	24 ft
Observation Well 1	16.1 ft
Observation Well 2	9.2 ft

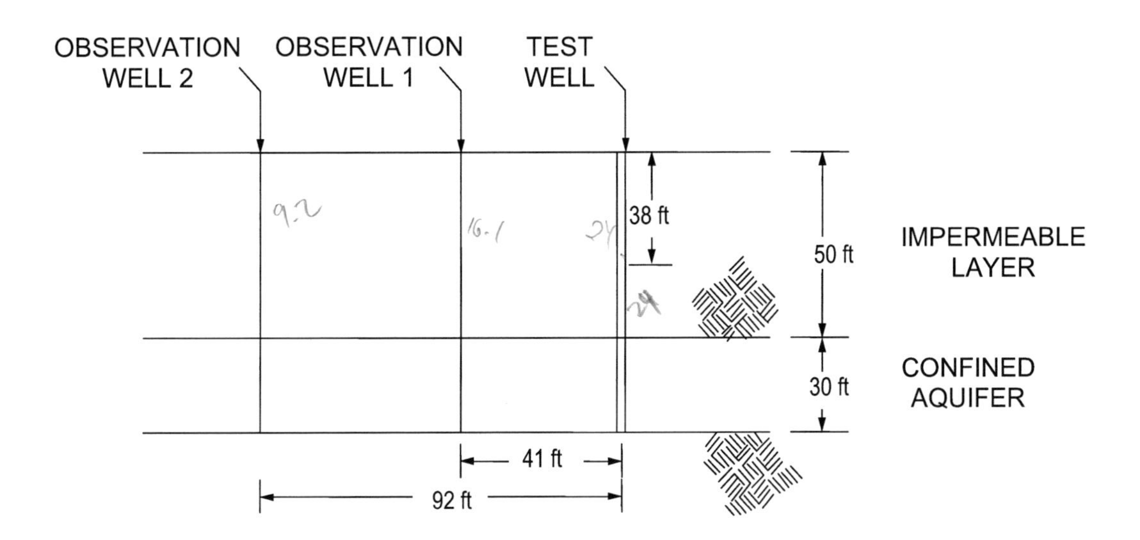

The permeability (fps) of the aquifer is most nearly:

(A) 1.9×10^{-4}
(B) 2.6×10^{-4}
(C) 3.2×10^{-4}
(D) 4.4×10^{-4}

123. A small community is building a water storage standpipe. The community wants to maintain a minimum system pressure of 30 psi and not exceed a maximum system pressure of 50 psi, while providing the largest useable storage volume for the standpipe. Assume 5-psi system losses and a 10-ft tank diameter. The high water level (ft) in the standpipe and the useable storage volume (gallons), respectively, are most nearly:

(A) 115.0; 27,000
(B) 115.0; 67,500
(C) 127.0; 27,000
(D) 127.0; 74,300

124. Ion exchange is used to soften well water with the characteristics given in the table below. The ion exchanger has leakage hardness of 20 mg/L as $CaCO_3$.

Raw Water Quality	
Parameter	**Concentration (mg/L)**
Ca^{+2}	98
Mg^{+2}	22
Na^{+}	12
HCO_3^{-}	320
SO_4^{-2}	65
Cl^{-}	21

To obtain a finished water hardness of 110 mg/L as $CaCO_3$, the percentage of water that must be bypassed and blended is most nearly:

(A) 26%
(B) 29%
(C) 33%
(D) 38%

125. Water flows into an old ductile-iron (C = 100) 12-in.-diameter distribution header at 5,000 gpm. A cross-over piping arrangement connects this header to an 8-in. auxiliary header with exit flows as shown in the figure below. Water demands at Points B, C, and D are 2,500, 1,000 and 1,500 gpm, respectively.

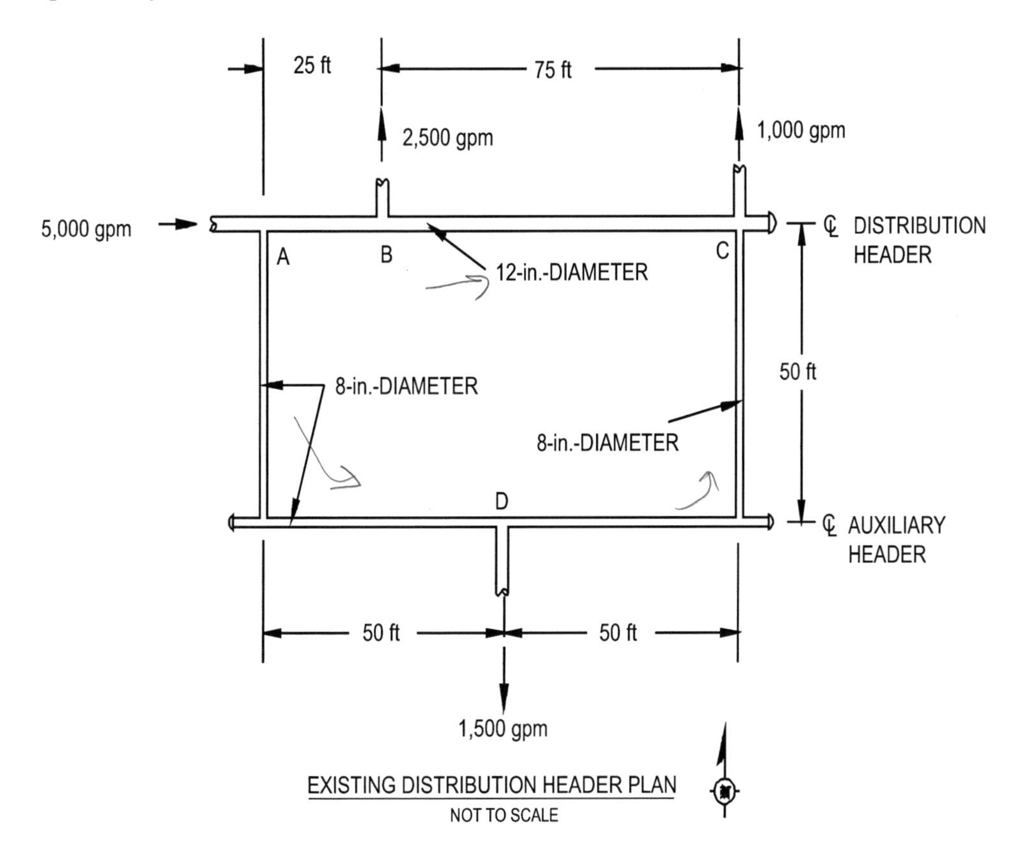

EXISTING DISTRIBUTION HEADER PLAN
NOT TO SCALE

For the purposes of this question, minor losses are to be neglected, and the initial flow distribution should be taken as listed in the following table.

Pipe	Flow Distribution (gpm)
AB	3,000
BC	500
CD	500
AD	2,000

After one iteration of the Hardy Cross method, the flow-correction value (gpm) that would be applied to the network is most nearly:

(A) 280
(B) 410
(C) 520
(D) 775

ENVIRONMENTAL MORNING SAMPLE QUESTIONS

126. The concentration of chlorine at the breakpoint indicates the presence of:

(A) oxidized chlorine
(B) free chlorine
(C) reduced chlorine
(D) chloramine

127. A 3.0-MGD surface water supply is treated with conventional filtration followed by chlorination. The chlorine dose is controlled to be not greater than 1.8 mg/L to avoid total trihalomethane problems. The chlorine demand is 45% of the dosage. The maximum free available chlorine concentration (mg/L) is most nearly:

(A) 0
(B) 0.81
(C) 0.99
(D) 1.80

128. The log inactivation of *Giardia* to meet the requirements of the surface water treatment rule is most nearly:

(A) The surface water treatment rule is not the regulation that establishes *Giardia* inactivation requirements.
(B) 2.0
(C) ln 10
(D) 3.0

129. Large floating masses of algae are a problem in many bodies of water. To control the algae, an effort must be made to control the limiting nutrient in algae growth. What is the limiting nutrient in the vast majority of lakes?

(A) Phosphorus
(B) Ammonia
(C) Nitrate
(D) Potassium

130. What two ions are responsible for nutrient enrichment?

(A) Nitrate and phosphate
(B) Calcium and phosphate
(C) Aluminum and nitrate
(D) Nitrate and sodium

ENVIRONMENTAL MORNING SAMPLE QUESTIONS

131. The most likely effect of the oxidized microzone of lake sediment on chromium would be to:

(A) promote release
(B) convert CrVI to CrIII
(C) increase algal uptake
(D) prevent release

132. If the bottom of a lake is anaerobic, what form of nitrogen would you find at the bottom of the lake?

(A) NO
(B) NO_3
(C) NH_3
(D) NH_2O

133. If 250 lb of algae are harvested from a lake, how may they be used?

(A) Saved until N:P ratio is upset and added to provide the appropriate ratio
(B) Used as a humic, N, and P amendment to soils
(C) Burned once dry
(D) Used for nutrients recovery

134. A wastewater treatment plant discharges to a receiving stream. After mixing of the wastewater effluent and the receiving stream the following data apply:

$CBOD_5$	= 6 mg/L		Wastewater temperature	=	24°C
BOD_5	= 8 mg/L		Upstream temperature	=	18°C
BOD_{ULT}	= 10 mg/L		Downstream temperature	=	20°C

Reaeration rate constant (20°C base e)	0.40 day^{-1}
Deoxygenation rate constant (20°C base e)	0.23 day^{-1}
Reaeration temperature correction coefficient	1.024
Deoxygenation temperature correction coefficient	1.047
Dissolved oxygen	6 mg/L
Stream velocity	1 fps

Assuming the time of travel required to achieve the maximum dissolved oxygen deficit in the stream is 3 days, the minimum dissolved oxygen concentration (mg/L) is most nearly:

(A) 2.9
(B) 5.2
(C) 6.2
(D) 7.3

Questions 135–136: A stationary engine is being used to generate electricity at an industrial site (**Figure 1**). The exhaust stack gas is being investigated and relevant information concerning pollutants in the exhaust gas are given in **Figures 2** and **3**.

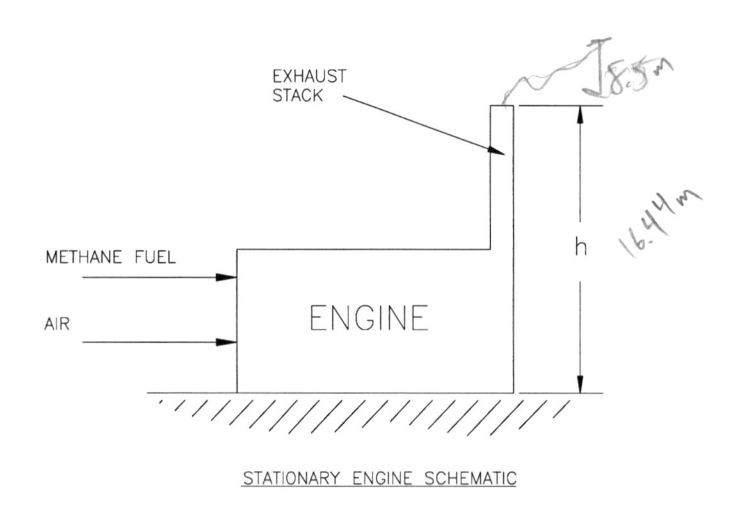

STATIONARY ENGINE SCHEMATIC

Data:

Exhaust stack height, h = 12 times stack diameter
Exhaust stack diameter, D = 1.37 m
Exhaust stack gas temperature = 500°C
Net Btu content of methane (CH_4) = 1,000 Btu/ft^3
Atomic weights: Carbon = 12.01; Hydrogen = 1.00
Fuel usage rate by cogeneration unit = 3,000 lb/hr
Plume rise = 8.5 m

Questions 135–136 (Continued)

Dispersion Formulae and Information:

σ_y and σ_z = Crosswind and vertical plume standard deviations in meters (from Figure 2)

$\bar{u}$ = Mean wind speed at the exhaust stack height = 4.2 m/s

h_e = Effective stack height in meters = h actual + Δh (plume rise)

g = Acceleration of gravity = 9.80 m/s^2

r = Stack radius in meters

T_s = Stack gas temperature, K = 273.16 + °C

T_a = Ambient air temperature = 287.33 K

V = Exhaust gas velocity at stack in m/s

F = Buoyancy flux = $gr^2v[1 - (T_a/T_s)]$

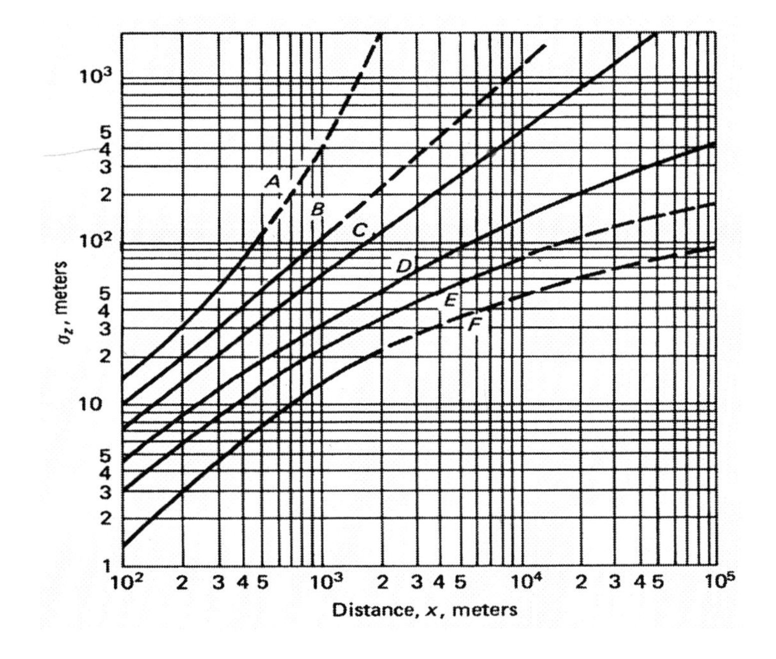

VERTICAL STANDARD DEVIATIONS OF A PLUME

FIGURE 1

GO ON TO THE NEXT PAGE

Questions 135–136 (Continued)

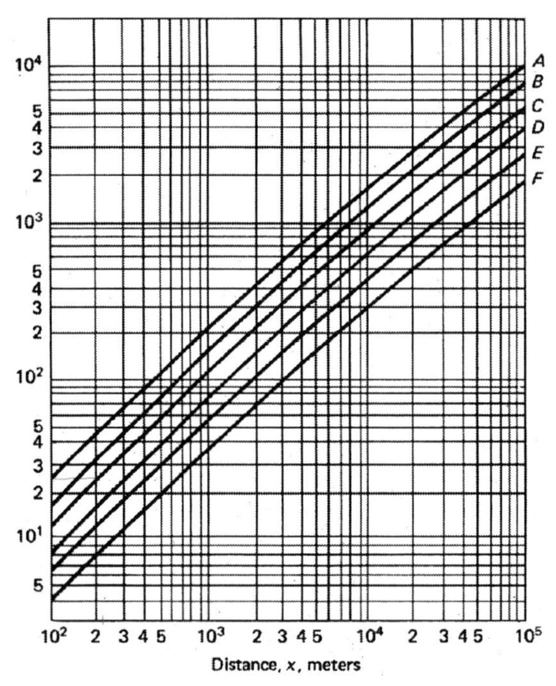

HORIZONTAL STANDARD DEVIATIONS OF A PLUME

A - EXTREMELY UNSTABLE
B - MODERATELY UNSTABLE
C - SLIGHTLY UNSTABLE
D - NEUTRAL
E - SLIGHTLY STABLE
F - MODERATELY STABLE

FIGURE 2

NOTE: Effective stack height shown on curves numerically.
SOURCE: Turner, O. B., "Workbook of Atmospheric Dispersion Estimates," Washington, DC, U.S. Environmental Protection Agency, 1970.

GO ON TO THE NEXT PAGE

Questions 135–136 (Continued)

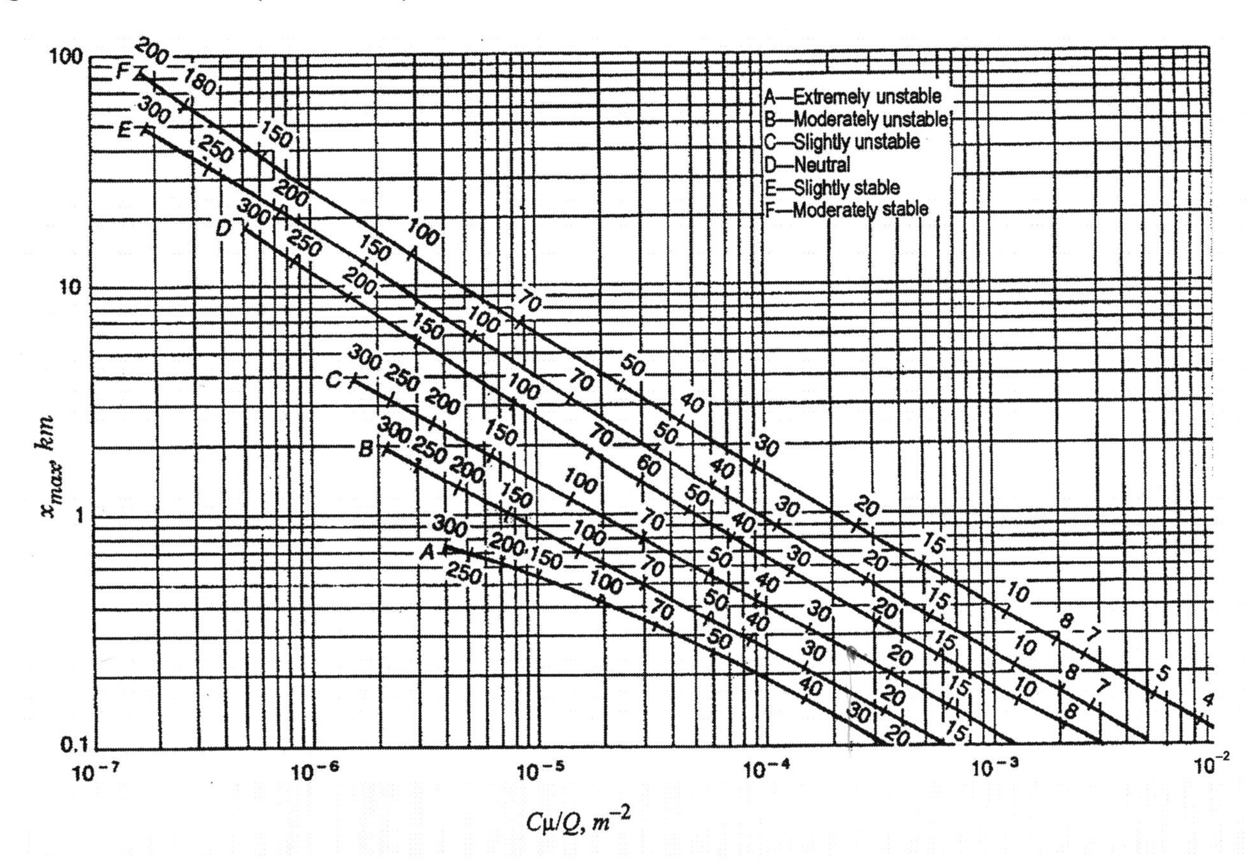

FIGURE 3

135. For a slightly unstable condition, the distance (meters) downwind to reach the maximum ground-level NO_x concentration is most nearly:

(A) 150
(B) 250
(C) 450
(D) 750

136. On an overcast day with a windspeed of 2.5 m/s, the stationary engine emits 0.15 lb $NO_x/10^6$ Btu. The maximum centerline NO_x concentration ($\mu g/m^3$) at 5 km downwind is most nearly:

(A) 2
(B) 5
(C) 7
(D) 10

137. What is the generally accepted method for measuring attainment with the National Ambient Air Quality Standard for nitrogen dioxide?

(A) Nondispersive infrared spectroscopy
(B) Calorimetric using Saltzman method
(C) Pararosaniline method
(D) Chemiluminescence method

138. Bioaerosol sampling methods are applicable to all of the following **EXCEPT** to:

(A) survey the types and concentrations of microorganisms
(B) measure allergens in residential and occupational settings
(C) examine microbiological air quality in food processing plants and animal containment buildings
(D) measure compliance with the TLV for total culturable organisms and particles

139. Federal air quality standards address which of the following contaminants?

(A) Nitrogen dioxide, ozone, particulates and biosolids
(B) Sulfur dioxide, particulates, airborne pathogens and arsenic
(C) Carbon monoxide, nitrogen dioxide, ozone and particulates
(D) Sulfur dioxide, oxygen free radicals, hydrocarbons and asbestos

140. The size and density of two particles are as follows:

Particle Size (μ)	Particle Density (g/cm^3)
10	1
5	4

The aerodynamic diameters (μ), respectively, of the particles are most nearly:

(A) 10 and 20
(B) 10 and 10
(C) 10 and 5
(D) None of the above

GO ON TO THE NEXT PAGE

141. A preliminary dispersion modeling analysis of NO_x emissions from gas-fir pipeline compressor station reveals that ambient air exceedences occur o' property line. To obtain an air quality permit for the station, the modeli… modeled exceedences. Which of the following options would be the best alte… eliminating the modeled exceedences?

(A) Increasing exhaust stack heights up to BEP
(B) Increase engine exhaust stack diameter
(C) Reduce the air/fuel ratio of engines
(D) Quench the discharge

142. The National Ambient Air Quality Standard (NAAQS) for NO_2 is 0.053 ppm (annual average). A dispersion modeling analysis of NO_x emissions from a source shows a maximum ambient receptor concentration of 220 $\mu g/m^3$. The receptor elevation is 6,500 ft (P_{bar} = 23.4 in. Hg), and the ambient temperature is 70°F. The NAAQS ($\mu g/m^3$) is most nearly:

(A) 50
(B) 80
(C) 200
(D) 450

143. The vapor pressure of mercury at 70°F and 760 mm Hg is 0.002 mm Hg. If the mercury were allowed to evaporate to equilibrium in an enclosed space, the theoretical concentration (ppm) in air would be most nearly:

(A) 2.6
(B) 31
(C) 52
(D) 13,000

$\frac{0.002}{760} \times 10^6 = 2.6$ ppm

144. A log normally distributed particle mass has a geometric standard deviation σ_g of 2.0 microns and a geometric mean diameter of 4 microns. A total of 84.1% of the particles are likely to have sizes (microns) less than:

(A) 2
(B) 4
(C) 6
(D) 8

A 500-MW power plant uses typical Pittsburgh seam coal at full load. The thermal efficiency is 35%. The following data are also given.

Coal heating value	13,600 Btu/lb
Emission factor (lb/ton):	
Particulate	139.2
SO_2	60.8

With a coal flow rate of 40 tons/hr, the particulate emission rate (tons/hr) is most nearly:

(A) 2.75
(B) 12.5
(C) 25.0
(D) 160

146. The Stokes' terminal velocity (m/s) of a cigarette smoke particle (density = 2,000 kg/m^3 and diameter = 5 μ) in air is most nearly:

(A) 6.05×10^{-3}
(B) 1.98×10^{-3}
(C) 1.50×10^{-3}
(D) 3.00×10^{-3}

147. A cyclone has the following dimensions:

Inlet duct height	=	0.6 m
Inlet duct width	=	0.15 m
Dust outlet diameter	=	0.3 m
Gas outlet diameter	=	0.4 m
Cyclone body length	=	1.5 m
Cyclone cone length	=	2.5 m

If the particle density is 1,950 kg/m^3 and the gas viscosity is 1.7×10^{-5} kg/(m•s), the effective number of turns is most nearly:

(A) 3.3
(B) 4.6
(C) 6.9
(D) 9.2

GO ON TO THE NEXT PAGE

148. A fan turning at 950 rpm supplies 725 cfm of air. The fan speed (rpm) required to move 1,000 cfm of air is most nearly:

(A) 690
(B) 950
(C) 1,100
(D) 1,310

149. The following information applies to the reaction of methane:

$$CH_4 + 2O_2 \rightarrow CO_2 + 2H_2O$$

Species	Enthalpy of Formation (J/mol)
CH_4	–74,980
O_2	0
CO_2	–394,088
H_2O	–242,174

The total heat of reaction (J/mol) is most nearly:

(A) 561,000
(B) 674,000
(C) 803,000
(D) 953,000

GO ON TO THE NEXT PAGE

150. A proposed air pollution control device has the following efficiency for the reported source particulate size distribution.

Particulate Size Range (μm)	Particle Size Distribution (% in size fraction)	Control device efficiency (%)
0–10	30	45
11–20	10	50
21–50	60	90

The collection efficiency for the device is most nearly:

(A) 50%
(B) 60%
(C) 70%
(D) 90%

GO ON TO THE NEXT PAGE

SAMPLE QUESTIONS FOR THE AFTERNOON PORTION OF THE EXAMINATION IN ENVIRONMENTAL ENGINEERING

Subject	Questions
Air	
Control Strategies	4
Solid, Hazardous, and Special Waste	
Municipal Solid Waste (MSW), Commercial, and Industrial Wastes	10
Hazardous Waste, Special, and Radioactive Waste	10
Environmental Assessments, Remediation, and Emergency Response	
Environmental Assessments	8
Remediation	8
Public Health and Safety	10
Total:	50

ENVIRONMENTAL AFTERNOON SAMPLE QUESTIONS

501. An incombustible porous product curing process involves two stages. In the first stage, a formed wet blanket is heated to dry the pack and cure the adhesive. In the second stage, the blanket is cooled to 150°F by 75°F air.

Data on the input material:

- A blanket 6 in. thick, 6 ft wide, at 75°F enters the first stage at 1 fps. The blanket has an effective 6.25% moisture content by weight. The blanket density is 1.6 pcf.
- The heat capacity of the adhesive solids and the blanket material is 2.0 Btu/(lb-°F).

In the curing stage, dry air at 75°F is heated by ventilating another process to:

- Evaporate the water from the input blanket.
- Raise the temperature of the blanket to 400°F to complete the cure.
- Reduce the blanket temperature to 225°F in a heat recovery section.

No significant particulate matter is produced.

The heating rate (Btu/hr) required to evaporate all moisture and raise the temperature of the blanket to 400°F is most nearly:

(A) 8,600,000
(B) 11,200,000
(C) 11,500,000
(D) 15,000,000

502. An incineration facility burning medical waste is required to report combustion efficiency. The flue gas has the following characteristics:

CO = 100 ppm @ 12% O_2 on a dry basis
CO_2 = 5,000 ppm @ 7% O_2 on a dry basis

With the above flue gas characteristics and using the EPA-required oxygen concentration on a dry basis (7% O_2), the combustion efficiency is most nearly:

(A) 94.5%
(B) 97.0%
(C) 98.0%
(D) 99.9%

503. The EPA new source performance standard (NSPS) for a large utility boiler is 0.2 lb NO_x (as NO_2) per 10^6 Btu heat input. The boiler burns natural gas. You have been hired by the utility to evaluate the following stack test data to determine if the boiler is in compliance.

Natural gas fuel rate	149.9×10^3 scfh
Ave NO_x conc.	117.27 ppmv
Stack velocity	28.84 fps
Stack diameter	92 in.
Stack pressure	23.56 in. Hg
Stack temperature	355°F
Moisture in stack gas	12.25%

1 ppm $NO_2 = 1.194 \times 10^{-7}$ lb NO_2/scf dry stack gas

Assume gas Btu value is 1,000 Btu/scf.

The lb NO_x per 10^6 Btu heat input is most nearly:

(A) 0.03
(B) 0.2
(C) 0.3
(D) 1.0

504. Which controls for criteria air pollutants are required for a major modification to an existing facility in an area that is in attainment with the national ambient air quality standards.

(A) best available control technology
(B) source pollution control guide
(C) lowest achievable emissions rate
(D) maximum achievable control technology

GO ON TO THE NEXT PAGE

ENVIRONMENTAL AFTERNOON SAMPLE QUESTIONS

Questions 505–507: A recycling program is to be instituted in a town that has a mass burn incinerator. The incinerator is capable of incinerating 600 tons/day (tpd). The town plans to recycle newspaper, corrugated board, glass, aluminum, and bi-metallic cans. The town's typical waste is:

Component	Waste Stream (%)	Heating Value (Btu/lb)
Newspaper	6.4	7,200
Corrugated board	9.0	7,000
Other paper	25.0	7,200
Plastic	7.4	14,000
Yard waste	19.6	2,800
Food wastes	20.1	1,500
Ferrous (Bi-metallic cans)	5.0	300
Aluminum	1.0	0
Glass	6.5	60
Total	100.0%	4,814

505. With the incinerator operating at 75% of capacity, the loss (Btu/day) to the incinerator with a recycling program efficiency of 50% is most nearly:

(A) 0.5×10^9
(B) 0.7×10^9
(C) 1.0×10^9
(D) 1.3×10^9

506. With a total daily waste collection rate of 500 tpd and a recycling efficiency of 40%, the daily revenue from glass at $30/ton and aluminum at $100/ton is most nearly:

(A) $400
(B) $600
(C) $1,500
(D) $1,750

507. At 450 kWh/ton of refuse incinerated, the annual gross revenue from electricity sales of this incinerator at 95% availability and sales of electricity at $0.04/kWh is most nearly:

(A) $10,300
(B) $156,000
(C) $3,745,000
(D) $4,125,000

Questions 508–509: A membrane-lined landfill leachate collection system consists of a sand drainage layer (hydraulic conductivity = 0.1 cm/s) tributary to a system of leachate collection pipes. The leachate collection pipes are perforated HDPE with a Manning's *n* coefficient of 0.009. The slope of the pipe is 1.2%. The peak daily lateral drainage rate in the leachate collection drainage layer is 1.5 in., and the peak infiltration rate into the sand drainage layer is 0.00002 cm/s.

508. The initial cell of the landfill is 10 acres. The peak leachate flow (cfs) for this cell is most nearly:

(A) 0.2
(B) 0.4
(C) 0.6
(D) 0.8

509. The peak flow of the second cell of the landfill is 0.45 cfs. The required leachate collection pipe nominal size (in.) is most nearly:

(A) 4
(B) 6
(C) 8
(D) 10

---

510. Benzene is biodegrading in petroleum contaminated soil at the rate of 0.2%/day, corrected for chemical dispersion. First order chemical kinetics is the only treatment technology. If the original contaminated soil concentration averages 17 mg/kg for benzene, the time (yrs) required to achieve a soil benzene concentration of 5 mg/kg, is most nearly:

(A) 0.5
(B) 1.12
(C) 1.67
(D) 2.33

511. Sources of leachate in a well-designed landfill cell can come from the following sources:

I. Precipitation
II. Water in the solid waste
III. Overland flow from other landfill cells
IV. Water entering the side of the landfill cell from the surrounding cells and area
V. Water entering the bottom of the landfill cell

(A) I and II only
(B) I, II, and III only
(C) I, II, III, and IV only
(D) I, II, III, IV, and V

512. Municipal wastewater sludges must have the following minimum percentage of solids content to be accepted for landfill disposal:

(A) 5%
(B) 10%
(C) 20%
(D) 50%

513. The minimum depth (in.) of a clay infiltration barrier in the final cover on a sanitary landfill is considered to be most nearly:

(A) 12
(B) 18
(C) 24
(D) 30

514. Landfill gas consists of 47.5% methane (CH_4) on a dry volume basis. Landfill gas has a heating value of 500 Btu/scf. Using a chemical separation process (i.e., removal of non-Btu gases), the product gas is 90% methane. The product gas is to be enriched with commercially available propane (2,350 Btu/scf) for commercial pipeline use (1,000 Btu/scf). The propane enrichment is most nearly:

(A) 4.0%
(B) 8.0%
(C) 12.0%
(D) 16.0%

Questions 515–517: A test burn waste mixture consisting of three designated principal organic hazardous constituents (POHCs) is incinerated at 1,000°C. The waste feed stream is as follows:

POHC	Inlet (kg/hr)	Outlet (kg/hr)
Chlorobenzene	153	0.010
Toluene	432	0.037
Xylene	435	0.070

515. With a required destruction and removal efficiency (DRE) of 99.99%, the test burn indicates that the:

(A) unit is in compliance with all POHCs
(B) chlorobenzene discharge is not in compliance
(C) toluene discharge is not in compliance
(D) xylene discharge is not in compliance

516. If all the chlorine in the feed is converted to hydrogen chloride (HCl), the HCl flow rate (kg/hr) prior to control is most nearly:

(A) 5
(B) 25
(C) 50
(D) 100

517. The maximum inlet flow rate (kg/hr) of chlorobenzene allowable to achieve an HCl discharge rate not to exceed 1.8 kg/hr is most nearly:

(A) 400
(B) 450
(C) 500
(D) 550

ENVIRONMENTAL AFTERNOON SAMPLE QUESTIONS

Questions 518–520: An RCRA Subtitle C hazardous waste landfill is planned with the cross section shown in the figure below.

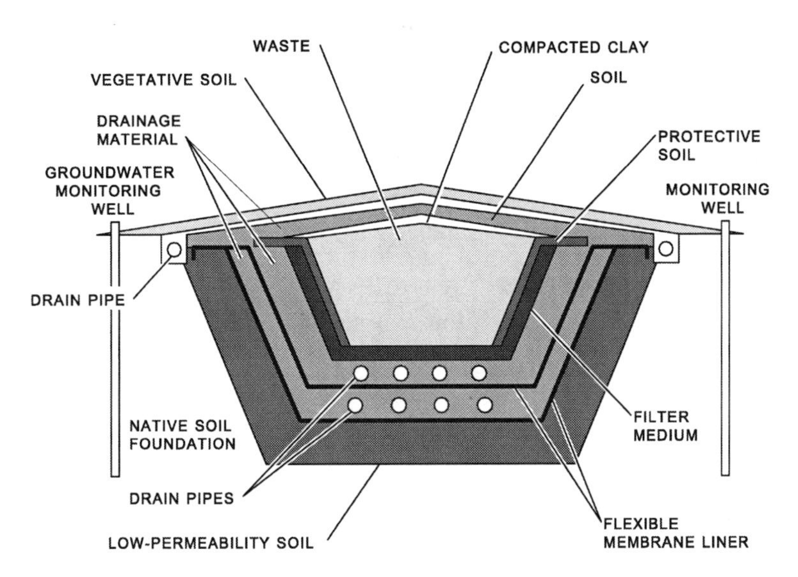

518. The EPA-recommended maximum slope (horizontal:vertical) for the flexible membrane liners is:

(A) 1:1
(B) 2:1
(C) 3:1
(D) 4:1

519. The "witness zone" is that section of the landfill between the:

(A) low-permeability soil and native soil foundation
(B) lower and upper flexible membrane liners
(C) upper flexible membrane liner and the filter medium
(D) waste and the upper compacted clay layer

520. Per EPA recommendations, the low-permeability soil should have an in-place conductivity (cm/s) less than:

(A) 1×10^{-4}
(B) 1×10^{-5}
(C) 1×10^{-6}
(D) 1×10^{-7}

Questions 521–523: A manufacturer generates 700 gallons/month of a liquid organic waste that contains 4%, by weight, of chlorobenzene as its principal organic hazardous component (POHC). The facility stores the liquids on site in tanks for 6 months. The specific gravity of the waste is 1.1, the pH is 7.9, and the flashpoint is 155°F.

521. The generator is considering less expensive options to incineration for management of the chlorobenzene. Options available under the federal RCRA generator standards are:

(A) disposal in an approved hazardous waste landfill
(B) on-site treatment to volatilize the chlorobenzene
(C) recycling the solution directly into the process from which it was generated
(D) (B) and (C)

522. Under the federal Resource Conservation and Recovery Act (RCRA), the generator classification and compliance status are:

(A) conditionally exempt small-quantity generator, in compliance with the regulation
(B) small-quantity generator, not in compliance
(C) large generator, in compliance
(D) large generator, not in compliance

523. During the incineration process, the 4,200 gallons are burned in a batch burn. The maximum allowable release of chlorobenzene (kg) in the batch is:

(A) 0.0375
(B) 0.070
(C) 0.15
(D) 0.30

GO ON TO THE NEXT PAGE

524. A subsurface remedial treatment technology costs $245,000 to construct initially with annual operation and maintenance costs of $9,000 for a 5-yr operational life. Using an annual interest rate of 7% and no equipment salvage value, the annualized cost for the remedial treatment technology is most nearly:

(A) $58,000
(B) $60,000
(C) $65,000
(D) $69,000

525. High molecular weight and strongly hydrophobic organic compounds are common soil contaminants due to petroleum crude oils spills. These compounds are most often partitioned to soils containing:

(A) silicate clay
(B) humic matter
(C) sand
(D) gravel

526. Transport of microorganisms to the groundwater depends mostly on:

(A) a combination of the amount of desiccation (cracks and fissures) and rainfall
(B) the amount of microorganisms applied
(C) the size of the microorganisms
(D) the pH of the soil

527. Which of the following geophysical measuring techniques would be used to assess the downhole vertical continuity of a clay stratum?

(A) terrain conductivity
(B) resistivity
(C) gamma ray logging
(D) ground penetrating radar

GO ON TO THE NEXT PAGE

528. A municipal landfill utilizes a compacted 1.08-m-thick clay liner that has a hydraulic conductivity of 1×10^{-7} cm/sec. If the depth of the leachate above the clay liner is 30 cm and the porosity of the clay is 55%, the time (years) required for the leachate to migrate through the liner is most nearly:

(A) 15
(B) 17
(C) 25
(D) 27

529. High molecular weight and strongly hydrophobic organic compounds are common soil contaminants (due to petroleum crude oils spills) but are not highly soluble in ground water. They do appear in groundwater monitoring samples in low concentrations. The most feasible mechanism for this occurrence is:

(A) a change in the electronic charge on clay particles
(B) the microbiological decomposition of humic matter
(C) colloidal transport in groundwater
(D) the decomposition of organic compounds

530. $LD_{50/30}$ refers to:

(A) the concentration in drinking water that is associated with the death of 50% of a population of 30-kg (STD) test animals
(B) the dose associated with the death of 50% of a population standardized to a 30-mg/(kg•day) dose rate equivalency
(C) the dose (via dermal exposure) that is associated with the death of 50% of a population of (STD) rats in 30 days
(D) the dose expected to cause the death of 50% of an exposed population within 30 ~~min~~ days

Questions 531–532: A bulk tanker truck carrying 5,000 gallons of methyl tertiary-butyl ether (MTBE), CASS#1634-04-4, has collided with a bridge structure over a small river near a community. The MTBE tank has ruptured, releasing an unknown amount of gasoline additive into the river about 1/4 mile upstream from the town's drinking water source intake.

The following MTBE data apply to this problem:

Oral R_fD	= 0.005 mg/(kg•day)
Inhalation R_fD	= 0.143 mg/(kg•day)
Manufacturer's suggested exposure limit*	= 600 mg/m^3 (or 200 ppm)
Vapor pressure @25°C	= 245 mm Hg
Density at 4°C	= 0.741 g/cm^3

*No TLV exists for MTBE

531. The Hazard Index (HI) criteria is often used for potential health risk associated with a chemical exposure incident. If the MTBE exposure intake is 0.00275 mg/(kg•day), the HI for an oral exposure in the drinking water system is most nearly:

(A) 0.11
(B) 0.19
(C) 0.37
(D) 0.55

532. It is proposed to treat the town's water supply to remove MTBE to an acceptable concentration. Assume a 365-day exposure duration and complete absorption into the body. The maximum MTBE concentration (mg/L) in the drinking water to protect a 15-kg child population consuming 1.0 L of water per day is most nearly:

(A) 0.00075
(B) 0.0075
(C) 0.075
(D) 0.75

533. Which of the following remediation technologies is most effective for removing high concentrations of low-solubility petroleum compounds from containment soil?

(A) Pump and treat wells
(B) Soil wash with a weak sulfuric acid/water mixture
(C) Injection of bacteria cultures
(D) Soil vapor extraction wells

534. An upgrade/expansion is proposed for the local POTW. Part of the proposal is to apply the sludge to farmland. The sites to which sludge will be applied have never previously received sludge. **Table 1** provides an estimate of nitrogen mineralization rates for sludges, and **Table 2** is an estimate of ammonia nitrogen availability coefficients.

TABLE 1

Sludge Type	Available Fraction of N (other than NH_3-N)
Lime stabilized	0.30
Aerobically digested	0.30
Anaerobically digested	0.20
Composted	0.10

TABLE 2

Method of Application	Sludge pH $\leq$ 10	Sludge pH > 10
Injection	0.95	0.95
Incorporated within 24 hr	0.85	0.75
Incorporated within 1–7 days	0.70	0.50
Incorporated after 7 days or no incorporation	0.50	0.25

Nitrogen is the limiting nutrient of concern. When designing the POTW expansion and coordinating sludge application with the farmers, which of the following practices should be considered to minimize land application acreage required by the POTW?

(A) Application of sludge to alfalfa or clover
(B) Lime stabilization and no incorporation
(C) Anaerobic digestion and incorporation within 1–7 days
(D) Aerobic digestion and injection

GO ON TO THE NEXT PAGE

535. A 40-ft-thick confined aquifer has a piezometric surface 85 ft above the bottom-confining layer. Groundwater is being extracted from a 4-in.-diameter fully penetrating well. The pumping rate is 35 gpm. The aquifer is relatively sandy with a hydraulic conductivity of 175 gpd/ft^2. Steady-state drawdown of 5 ft is observed in a monitoring well 10 ft from the pumping well. The drawdown (ft) in the pumping well is most nearly:

(A) 9
(B) 14
(C) 19
(D) 24

536. A landfill cell has an average width of 200 ft and an average length of 1,000 ft. The landfill cell is 15 ft below grade and 20 ft above grade. Cover represents 15% of the total air space.

The added cost for suitable fill for cover is $100,000. An alternate cover method is proposed using a reusable cover. The cost for this cover and replacements throughout the life of the landfill cell is $250,000. If the revenue from the added space is estimated to be $25/yd^3, the savings achieved using the reusable cover are most nearly:

(A) $350,000
(B) $560,000
(C) $725,000
(D) $825,000

537. A mixture of hazardous waste and a solid (nonhazardous) waste is:

(A) not considered a hazardous waste if it only contains *de minimus* amount of hazardous waste
(B) not considered hazardous waste if the mixture no longer retains the characteristic that made it hazardous; e.g., ignitability
(C) not considered a hazardous waste under any circumstances
(D) always considered a hazardous waste

GO ON TO THE NEXT PAGE

538. The incinerator shown in the figure below is designed to treat a chlorobenzene (C_6H_5Cl) contaminated gas stream from a manufacturing process. The incinerator burns natural gas as its auxiliary fuel. Air for the combustion process is provided through a preheater recovering heat from the furnace exhaust. The thermal efficiency of the incinerator is 85%. The heating values of chlorobenzene and natural gas are 17,000 and 24,000 Btu/lb, respectively. Saturation pressures for steam are tabulated below.

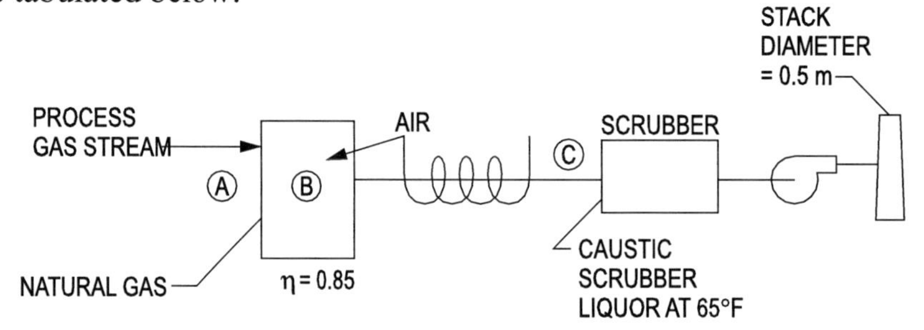

<u>Point A:</u>

Flow rate = 2,000 m^3/hr
Density = 0.74 kg/m^3
T = 300 F
P = 780 mm Hg
Enthalpy = 60 Btu/lb
Chlorobenzene = 1,200 ppm (by volume)
Remainder of gas is air with sufficient oxygen to burn the chlorobenzene.

<u>Point C:</u>

T = 600 F
P = 800 mm Hg
Density = 0.6 kg/m^3
Enthalpy = 136 Btu/lb

538. (Continued)

Saturation Pressures

Press. Lbf. Sq.In. p	Temp. Fahr. t	Specific Volume Sat. Liquid v_f	Specific Volume Sat. Vapor v_g	Internal Energy Sat. Liquid u_f	Internal Energy Evap. u_{fg}	Internal Energy Sat. Vapor u_g	Enthalpy Sat. Liquid h_f	Enthalpy Evap. h_{fg}	Enthalpy Sat. Vapor h_g	Entropy Sat. Liquid s_f	Entropy Evap. s_{fg}	Entropy Sat. Vapor s_g
14.696	211.99	.016715	26.80	180.10	897.5	1077.6	180.15	970.4	1150.5	.31212	1.4446	1.7567
15	213.03	.016723	26.29	181.14	896.8	1077.9	181.19	969.7	1150.9	.31367	1.4414	1.7551
16	216.31	.016746	24.75	184.45	894.4	1078.8	184.50	967.6	1152.1	.31858	1.4313	1.7499
17	219.43	.016768	23.39	187.59	892.1	1079.7	187.65	965.6	1153.3	.32322	1.4218	1.7450
18	222.40	.016789	22.17	190.59	889.9	1080.5	190.64	963.7	1154.4	.32762	1.4128	1.7404
19	225.24	.016810	21.08	193.45	887.8	1081.3	193.51	961.9	1155.4	.33181	1.4043	1.7361

Reference: Steam Tables by Joseph H. Keenan, F.G. Keyes, P.G. Hill and J.G. Moore.

If this incinerator operates as a treatment storage disposal facility under the Federal Resource Conservation and Recovery Act (RCRA), the maximum discharge rate (g/hr) of the chlorobenzene is most nearly:

(A) 0.1
(B) 1
(C) 100
(D) 1,000

GO ON TO THE NEXT PAGE

Questions 539–540: A 1,250-ft^3, 8-ft-diameter bed of spent activated carbon requires disposal. The carbon has been used in the removal of non-volatile impurities from a methyl tertiary-butyl ether (MTBE) production stream. Dry nitrogen gas will be used to reduce the MTBE concentration to the required spent bed disposal concentration of 500 parts per million (ppm), by weight. A spent bed sample was dried in the laboratory by blowing 25°C nitrogen gas over a 10-cm × 10-cm tray having a depth of 2.15 cm. The following drying data were obtained:

Time (min)	Sample Wt (g)	Time (min)	Sample Wt (g)
0	252.6	90	238.4
15	250.1	120	237.1
30	247.5	240	236.6
45	245.1	300	236.3
60	242.6	360	236.2
75	240.1	420	236.2

MTBE is the only adsorbed volatile compound, and it is completely purged at the constant weight point.

539. By raising the temperature of the nitrogen heating gas to 120°C, which of the following effects would **NOT** occur?

(A) MTBE vapor pressure would be higher.
(B) Less nitrogen gas would be required.
(C) The drying period would be shorter.
(D) More MTBE would be removed.

540. If 75,000 pounds of spent bed material were to be dried in a manner similar to that used in the laboratory test, the bed material would have to be applied over an area (ft^2) of most nearly:

(A) 3,880
(B) 5,175
(C) 7,760
(D) 14,510

ENVIRONMENTAL AFTERNOON SAMPLE QUESTIONS

Questions 541–543: An abandoned underground storage tank (UST) has been discovered during environmental assessment of a property. The buyer wants to remove the UST, but first it must be cleaned. Visual inspection through an access hatch reveals:

- The UST appears to be 8 ft in diameter × 12 ft long. The 12-ft dimension is horizontal.
- About 12 in. of sludge lies in the tank bottom, covered by about 2 in. of liquid.
- Tree leaves are visible, floating on the liquid, and in the muck under the liquid.
- Sample analysis reveals the liquid to be hydrocarbon-contaminated water at a pH of about 5.

Workers must enter the UST to remove the muck and sludge.

541. To qualify this UST as a "confined space," it must be established that it has:

(A) potential for temperature extremes, engulfment hazards
(B) limited access and/or toxic atmospheres
(C) potential for hazardous atmospheres
(D) potential for IDLH atmospheres

542. The respiratory protection equipment required for initial entry to remove sludge that is contaminated with an unknown hydrocarbon is:

(A) SCSR (self-contained self-rescuer)
(B) a filter mask with organic vapor cartridge
(C) SCBA operated in pressure demand mode
(D) SCUBA with anti-fog mark

543. In a chemical emergency, frequent atmospheric oxygen measurements are needed to specify the level of protective clothing. To recommend the proper respiratory protection equipment, an oxygen-deficient atmosphere (% oxygen by volume) as defined by NIOSH/OSHA is:

(A) 20.9% or less
(B) 19.5% or less
(C) 17.0% or less
(D) 15.3% or less

544. A worker is exposed to the following air concentrations of a contaminant:

Time	Contaminant Level (mg/m^3)
0800–1200	15
1200–1300	0
1300–1600	7
1600–2000	9

The 8-hr time-weighted average (TWA) exposure (mg/m^3) is most nearly:

(A) 7.8
(B) 9.7
(C) 10.6
(D) 14.6

545. If one HVL (half value) is 0.33 cm of lead, the thickness of lead (cm) required to reduce the dose rate from 10 mrem/hr to 2.5 mrem/hr is most nearly:

(A) 0.04
(B) 0.33
(C) 0.52
(D) 0.66

546. The metabolic process for expelling methyl mercury from the body follows first order kinetics, as expressed by the following equation:

$$C = C_o e^{-kt}$$

Assuming that the average excretion rate is 1.75% of the total body burden per day, the half-life (days) of this chemical in the body is most nearly:

(A) 40
(B) 50
(C) 60
(D) 70

547. A noise level of 100 dBA exists for 5 min followed by 50 min of a 60-dBA noise. The equivalent continuous constant noise level (dBA) for the entire 55-min period is most nearly:

(A) 80
(B) 90
(C) 100
(D) 160

548. A hydrocarbon has spilled in a room that has a volume of 150 m^3. The concentration of the hydrocarbon in the room air is 100,000 mg/m^3. Assume no additional evaporation or contaminant generation. Room air is evacuated using a fan that exhausts 1,000 cfm or 28.32 m^3/min (assume 100% outside makeup air). A concentration of 5 mg/m^3 is recommended before emergency responders can enter the room. Using a safety factor of 4, the estimated time (min) that elapses before responders can enter the room is most nearly:

(A) 60
(B) 120
(C) 180
(D) 210

549. One quart of hydrocarbon (MW = 92, Specific gravity = 0.87, and VP = 21 mm Hg @ 70°F) is spilled in a 20-ft × 30-ft × 10-ft room that is at 70°F. The maximum concentration (mg/m^3) of hydrocarbon in the room air is most nearly:

(A) 1,235
(B) 4,640
(C) 27,600
(D) 103,971

GO ON TO THE NEXT PAGE

550. An MTBE manufacturer's test data indicate breakthrough times for the following materials:

Less than 1 hr for:
Neoprene
Polyvinyl chloride
Cotton

Greater than 4 hr for:
Polyvinyl alcohol
Nitrile

Which personal protective clothing is the best chemical-resistant material to protect emergency response personnel from direct human contact with MTBE during the cleanup operation of a spill into a river?

(A) Nitrile
(B) Neoprene
(C) Cotton
(D) Polyvinyl alcohol

SOLUTIONS FOR SAMPLE QUESTIONS IN THE MORNING PORTION OF THE EXAMINATION IN ENVIRONMENTAL ENGINEERING

ENVIRONMENTAL MORNING SAMPLE SOLUTIONS

Answer Key for
Morning Sample Questions

101	A	**126**	B
102	D	**127**	C
103	D	**128**	D
104	D	**129**	A
105	A	**130**	A
106	A	**131**	A
107	A	**132**	C
108	A	**133**	B
109	B	**134**	C
110	A	**135**	B
111	C	**136**	B
112	C	**137**	D
113	B	**138**	D
114	A	**139**	C
115	D	**140**	B
116	B	**141**	A
117	D	**142**	B
118	D	**143**	A
119	A	**144**	D
120	C	**145**	A
121	A	**146**	C
122	A	**147**	B
123	C	**148**	D
124	B	**149**	C
125	D	**150**	C

ENVIRONMENTAL MORNING SAMPLE SOLUTIONS

101. In order to determine loading on the RBC process, it is first necessary to evaluate trickling filter performance.

Using the NRC equation (Metcalf & Eddy, 3rd Ed., 1991, *Wastewater Engineering, Treatment, Disposal, Reuse,* p. 409.)

$$E = \frac{100}{1 + 0.0561\sqrt{W/VF}}$$

E = Efficiency of BOD_5 removal across filter including clarifier (%)

W = BOD_5 influent loading to filter (lb/day)

V = Volume of filter (1,000 ft^3)

R = Recirculation ratio = Q_R/Q = 0.5

F = Recirculation factor

$$= \frac{1+R}{(1+R/10)^2} = \frac{1+0.5}{(1+0.05)^2} = 1.36$$

W = (1.5 MGD)(250 mg/L)(8.34)

= 3,127.5 lb BOD_5/day

$$V = \left(\frac{\pi D^2}{4}\right) \times \text{Depth} = \frac{\pi(75\text{ ft})^2}{4}(6\text{ ft})$$

$= 26{,}500\ ft^3$ or $26.5 \times 10^3\ ft^3$

$$E = \frac{100}{1 + 0.0561\sqrt{3{,}127/(26.5 \times 1.36)}} = 65.6\%$$

$\therefore$ Output from trickling filter = 3,127.5 (1–0.656)

= 1,075 lb/day

THE CORRECT ANSWER IS: (A)

102. Determine the efficiency of BOD_5 removal.

e = 40 + (92)(1.0 – 0.4) = 95%

THE CORRECT ANSWER IS: (D)

103. Metals bioaccumulate and regulations restrict application of metals to land in order to keep metals out of the human food chain.

THE CORRECT ANSWER IS: (D)

104. BOD, TSS, and metals can easily be collected as a composite and are stable in a refrigerated sampler. Due to difficulty in collecting a composite oil and grease sample, it is specified to be a grab sample.

THE CORRECT ANSWER IS: (D)

105. THE CORRECT ANSWER IS: (A)

106. Determine depth of flow in sewer for conditions given:

$$V_{full} = \frac{1.486}{0.013}(1/4)^{2/3}(0.004)^{1/2}$$

$$= 2.87 \text{ fps}$$

Self-cleansing V = 2 fps

$$\therefore \frac{V}{V_{full}} = \frac{2}{2.87} = 0.697 \cong 0.7$$

Refer to nomograph for hydraulic elements (ASCE, MOP 37, *Design & Construction of Sanitary & Storm Sewers*, 1969, pg. 87)

$$\text{If } \frac{V}{V_{full}} = 0.7 \qquad \text{then } \frac{d}{D} = 0.39$$

$$\therefore d = (0.39)(12) = 4.68 \text{ in.}$$

THE CORRECT ANSWER IS: (A)

107. Calculate the temperature in the stream:

$$\text{Temp} = \frac{4\ \text{MGD} \times \dfrac{1.547\ \text{cfs}}{\text{MGD}} \times 24°\ \text{C} + 18\ \text{cfs} \times 27°\ \text{C}}{4\ \text{MGD} \times \dfrac{1.547\ \text{cfs}}{\text{MGD}} + 18\ \text{cfs}} = 26.2°\ \text{C}$$

Correct the rate constants:

$k_d = 0.23(1.047)^{26.2-20} = 0.305$

$k_r = 0.4(1.024)^{26.2-20} = 0.46$

Calculate BOD_5 of mixture (effluent and stream):

$$BOD_5 = \frac{\dfrac{20\ \text{mg}}{\text{L}} \times 4\ \text{MGD} \times 1.547 + 18\ \text{cfs} \times \dfrac{4\ \text{mg}}{\text{L}}}{4\ \text{MGD} \times 1.547 + 18\ \text{cfs}} = 8.09\ \text{mg/L}$$

Calculate BOD_L of mixture:

$BOD_5 = BOD_L(1 - e^{-kt}) = 8.09 = BOD_L(1 - e^{-0.23(5)})$

$BOD_L = 11.76$ mg/L

Calculate the dissolved oxygen deficit:

$$D = \frac{k_d}{k_r}\left(L \times e^{-kt}\right) = \frac{0.305}{0.46}\left(11.76 \times e^{-0.305(3)}\right) = 3.12\ \text{mg/L}$$

THE CORRECT ANSWER IS: (A)

108. Calculate cadmium criteria for long term ecological protection of aquatic life.

$X = \exp[M_c(\ln(\text{hardness})) + b_c]$

where $M_c = 0.785$
Hardness = 125
$b_c = -3.49$

$X = e^{(0.785(\ln 125) - 3.49)}$

$X = 1.35\ \mu g/L$

THE CORRECT ANSWER IS: (A)

109. Calculate % solids and % volatile solids.

First net the dish weight from the data.

Wet sample	133.61 – 19.15 =	114.46
Dry sample	25.72 – 19.15 =	6.57
Ash	20.37 – 19.15 =	1.22

% Solids = dry/wet = 6.57 / 114.46 = 5.7%

% Volatile = (dry – ash)/dry = (6.57 – 1.22) / 6.57 = 81.4%

THE CORRECT ANSWER IS: (B)

110. Find first order reaction rate constant.

In – Out = Generated = Total

$$0 - 0 + Vr_c = V\frac{dc}{dt}$$

$$-KC = \frac{dc}{dt}$$

$$\int_{c_o}^{c} \frac{dc}{dt} = \int_{o}^{t} -Kdt$$

$$\ln C/C_o = -Kt$$

$$K = (\ln C/C_o) / (-T)$$

$$= \ln(10/200) / (-1)$$

$$= 3.0 \text{ hr}^{-1}$$

THE CORRECT ANSWER IS: (A)

111. Perform present worth analysis.

$$\text{PV/FA 6\%, 10 yr} = \frac{1-\frac{1}{(1+i)^n}}{i} = \frac{1-\frac{1}{(1.06)^{10}}}{0.06} = 7.360$$

<u>Chlorination/Dechlorination</u>

Labor: 30 hr/mo × 12 × \$20	\$7,200
Power: $\frac{1,500}{1,000}\text{kW}/\text{hr}\times 24\times 365\times \$0.07$	\$920
Hypochlorite: 100 gpd × 365 × \$0.87	\$31,755
Bisulfite: $10\text{ gpd}\times 365\times \frac{\$192.5}{55}$	\$12,775
Misc:	\$3,000
Annual OM&R:	\$55,650
Total PW: 600,000 + 55,650(7.36) =	\$1,009,584

<u>UV</u>

Labor: 50 hr/mo × 12 × \$20	\$12,000
Power: $\frac{(300)(75)}{1,000}\text{kW}/\text{hr}\times 24\times 365\times \$0.07$	\$13,797
Lamp Replacement: 50% × 300 × \$40	\$6,000
Misc:	\$3,000
Annual OM&R:	\$34,797
Total PW: 750,000 + 34,797(7.36) =	\$1,006,106

The UV system is calculated to be \$3,478 cheaper. However, planning level capital costs are at best accurate within 10%. OM&R projections are at best accurate to 5%. Thus, the UV system has a variability of 10% of 750,000 plus 5% of (34,797) (7.63) or \$87,805. The two alternatives are equal in cost within the accuracy of the analysis.

THE CORRECT ANSWER IS: (C)

112. Find critical depth

$$Q = \sqrt{\frac{gA^3}{B}}$$

where Q = 2 cfs
g = 32.3 ft/sec^2
A = area
B = channel width @ surfaces

$$A = 2(1/2d \times 4d) = 4d^2$$

$$B = 2 \times 4d = 8d$$

$$Q = \sqrt{\frac{g(4d^2)^3}{8d}}$$

$$Q^2 = \frac{g(64d^6)}{8d} = 8gd^5$$

$$d = \sqrt[5]{\frac{Q^2}{8g}} = \sqrt[5]{\frac{2^2}{8(32.2)}} = 0.435 \text{ ft}$$

THE CORRECT ANSWER IS: (C)

113. Reference: Bedient, Philip B. and Huber, Wayne C., *Hydrology and Flood Plain Analysis*, 2nd ed., Addison-Wesley, pg. 129

Area	Land Use	Area (Acres)	Fraction of Area	NRCS/SCS Runoff Curve Number (CN)	Fraction of Area × CN
A	Forest	8	0.2	55	11.0
B	Forest	12	0.3	55	16.5
C	Park/Open Space	6	0.15	61	9.2
D	Residential	6	0.15	75	11.3
E	Residential	4	0.1	75	7.5
F	Residential	4	0.1	85	8.5
		40.0	1.0		64.0

THE CORRECT ANSWER IS: (B)

114. The maximum overland flow time to Point 3 is determined by evaluating the times as follows:

(A to Pt 1) + (Pt 1 to Pt 2) + (Pt 2 to Pt 3) = 5 + 2 + 3 = 10
(B to Pt 1) + (Pt 1 to Pt 2) + (Pt 2 to Pt 3) = 6 + 2 + 3 = 11
(C to Pt 2) + (Pt 2 to Pt 3) = 10 + 3 = 13
(D to Pt 2) + (Pt 2 to Pt 3) = 12 + 3 = 15
(E to Pt 3) = 3
(F to Pt 3) = 4

The maximum overland flow time is 15 minutes from Area D.

For a 100-year frequency storm and a duration of 15 minutes, the design rainfall intensity is 2.1 in./hr.

THE CORRECT ANSWER IS: (A)

115. $V = \frac{1.49}{n} R^{2/3} s^{1/2}$

The velocity is a function of roughness (n) and slope (s). The hydraulic radius (R) is defined as follows:

$$R = \frac{\text{Area}}{\text{Wetted perimeter}}$$

For the rectangular channel

$$R = \frac{wd}{2d + w}$$

Thus R is **NOT** constant as area is varied (changes in d will cause changes in R). Therefore, velocity depends on area.

Velocity is a function of roughness, slope, and area (I, II, and III).

THE CORRECT ANSWER IS: (D)

116. Determine the unit hydrograph discharge after 10 hours.

The procedure for determining the unit hydrograph follows the method of Gupta (1989, pp. 302–304 and 310–311). The flow data indicate zero discharge prior to the rainfall event. The flow returns to zero after 14 hours. Based on these zero flows, it may be concluded that the base flow is zero. The total runoff volume (V) is obtained by integration of the flow vs. time response. The unit hydrograph for an event of equal duration is obtained by division of the measured flow values by the runoff depth, defined as the runoff volume divided by the drainage area.

Time (hr)	Runoff (cfs)	Avg. Runoff (cfs)	Duration (hr)	Volume (cfs-hr)
0	0.0			
2	2.5	1.25	2	2.50
4	5.0	3.75	2	7.50
6	7.5	6.25	2	12.50
8	4.0	5.75	2	11.50
10	2.0	3.00	2	6.00
12	1.0	1.50	2	3.00
14	0.0	0.50	2	1.00
Total				44.00

Runoff Depth $= \frac{(44 \text{ cfs-hr})(3{,}600 \text{ sec/hr})}{(40 \text{ acres})(43{,}560 \text{ ft}^2/\text{acre})} = 0.091 \text{ ft} = 1.1 \text{ in.}$

The measured discharge after 10 hours = 2 cfs. For a unit hydrograph (1-in. event), the corresponding discharge (Q) is calculated as:

Q = (2 cfs)(1 in.) / (1.1 in.) = 1.8 cfs

THE CORRECT ANSWER IS: (B)

117. Scope of Part I Group Stormwater Applications

Part I Group Applications asked "Who?" and "What?". "Where?" and "How Much?" were covered by Part II.

THE CORRECT ANSWER IS: (D)

118. Average Flow: 13 MGD × 8.34 × 11.5 mg/L = 1,246.8 lb/day × day/24 hr = 51.9 lb/hr

Peak flow: 32.5 MGD × 8.34 × 11.5 mg/L = 3,117 lb/day × day/24 hr = 129 lb/hr

THE CORRECT ANSWER IS: (D)

119. Convert the cation and anion concentrations to meq/L.

Cations:

Calcium: (51 mg/L) / (20 mg/meq) = 2.55 meq/L

Magnesium: (12 mg/L) / (12.1 mg/meq) = 0.99 meq/L

Sodium: (25 mg/L) / (23 mg/meq) = 1.09 meq/L

Sum of Cations = 4.63 meq/L

Anions:

Sulfate: (65 mg/L) / (48 mg/meq) = 1.35 meq/L

Chloride (25 mg/L) / (35.5 mg/meq) = 0.70 meq/L

Nitrate as N: (14 mg/L) / (14 mg/meq) = 1.00 meq/L

Bicarbonate: (84 mg/L) / (50 mg/meq) = 1.68 meq/L

Sum of anions = 4.73 meq/L

4.73 meq/L – 4.63 meq/L = 0.10 meq/L

Note: Fluoride can be omitted since the concentration is < 0.1 meq/L

THE CORRECT ANSWER IS: (A)

120. Determine the flow in the existing 10-in. pipe system. Use the Hazen-Williams equation and calculate the total dynamic head for a nominal 10-in. pipe diameter:

$Q = V A = 1.318\ C\ A\ R^{0.63}\ S^{0.54}$

$C = 100$

$A = (1/4)\pi\ D^2 = (1/4)\ (3.14159)\ (0.8333)^2 = 0.5453\ ft^2$

$R = D/4 = (0.8333/4) = 0.2083\ ft$

$$S^{0.54} = \frac{Q}{1.318\ C\ A\ R^{0.63}} = \frac{Q}{1.318(100)(0.5453)(0.2083)^{0.63}} = 0.03738\ Q$$

$S = 0.002273\ Q^{1.8519}$

$h_f = (3{,}000\ ft)\ (0.002273)\ Q^{1.8519}$

$TDH = \Delta H + h_f + V^2/2g$

$\Delta H = 560 - 530 = 30\ ft$

Q (gpm)	Q (cfs)	h_f (ft)	V = Q/A (fps)	$V^2/2g$ (ft)	ΔH (ft)	TDH (ft)
400	0.891	5.511	1.634	0.041	30.00	35.6
600	1.337	11.676	2.451	0.093	30.00	41.8
800	1.783	19.892	3.269	0.166	30.00	50.1
900	2.005	24.740	3.677	0.210	30.00	55.0
1,000	2.228	30.071	4.086	0.259	30.00	60.3
1,200	2.674	42.148	4.903	0.373	30.00	72.5

Plot the system head curve (TDH). The point of operation is where the system head curve crosses the pump curve.

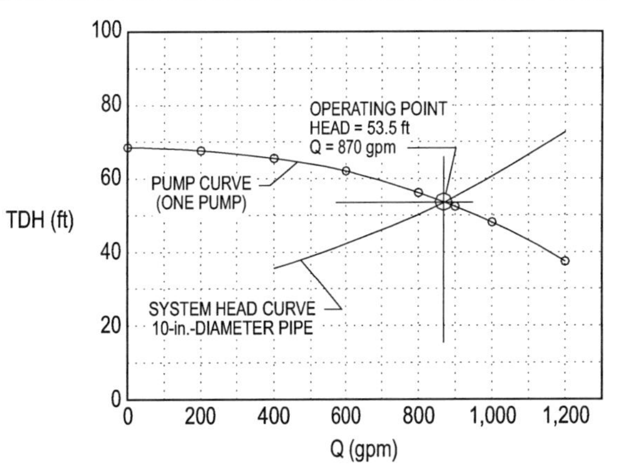

Q = 870 gpm

THE CORRECT ANSWER IS: (C)

GO ON TO THE NEXT PAGE

121. The following actions will reduce the tendency for pump cavitation:

II. Lowering the pump elevation
III. Increasing the suction diameter

THE CORRECT ANSWER IS: (A)

122. Determine the permeability of the confined aquifer

$$K = \frac{Q}{2\pi b(h_2 - h_1)} \ln\left(\frac{r_2}{r_1}\right)$$

Datum:
$h_o = 80 - 38 = 42$ ft
$h_w = 42 - 24 = 18$ ft
$h_1 = 42 - 16.1 = 25.9$ ft
$h_2 = 42 - 9.2 = 32.8$ ft

$$K = \frac{0.3 \text{ cfs}}{2\pi(30)(32.8 - 25.9)} \ln\left(\frac{92}{41}\right)$$

$$= \frac{0.3}{1{,}300.6}(0.808)$$

$$= 1.86 \times 10^{-4} \text{ fps}$$

THE CORRECT ANSWER IS: (A)

123. $30 \text{ psi} \times \frac{2.31 \text{ ft}}{\text{psi}} = 69.3 \text{ ft}$

$50 \text{ psi} \times \frac{2.31 \text{ ft}}{\text{psi}} = 115.5 \text{ ft}$

$5 \text{ psi} \times \frac{2.31 \text{ ft}}{\text{psi}} = 11.55 \text{ ft}$

Max height = 115.5 ft + 11.55 ft = 127.05 ft

$$\text{Volume} = \frac{\pi(10)^2}{4} \times (115.5 - 69.3) = 3{,}628.5 \text{ ft}^3$$

$3{,}628.5 \text{ ft}^3 \times 7.48 \text{ gal/ft}^3 = 27{,}141 \text{ gal}$

THE CORRECT ANSWER IS: (C)

124. $X = \dfrac{H_{Finish} - H_{Leak}}{H_{Raw} - H_{Leak}} = \text{Bypass Fraction}$

H_{Raw} Ca^{+2} = 98 mg/L ≈ 4.9 meq/L

Mg^{+2} = 22 mg/L ≈ 1.8 meq/L

Σ 6.7 meq/L × 50 = 335 mg/L as $CaCO_3$

$X = \dfrac{110 - 20}{335 - 20} \approx 29\%$

THE CORRECT ANSWER IS: (B)

125. Determine headloss equations for each pipe using the Hazen-Williams equation.

$Q = 0.281\ C\ D^{2.63}\ S^{0.54}$

$HL = S\ L = \dfrac{10.47\ L}{C^{1.85} D^{4.87}} Q^{1.85} = K\ Q^{1.85}$

where

- Q = flow (gpm)
- C = Hazen-Williams coefficient (100 for old cast iron)
- D = pipe diameter
- S = slope of energy grade line
- L = pipe length
- HL = headloss
- K = constant for each pipe

Determine K for each pipe

Pipe	C	D	L	K
AB	100	12	25	2.899×10^{-7}
BC	100	12	75	8.698×10^{-7}
CD	100	8	100	8.354×10^{-6}
AD	100	8	100	8.354×10^{-6}

125. (Continued)
Assume CW as the positive direction, and complete one iteration of the Hardy Cross method.

Pipe	Q (gpm)	HL (ft)	HL/Q (ft/gpm)	Corrected Q (gpm)
AB	+3,000	0.7851	0.000262	+3,785
BC	+500	0.0856	0.000171	+1,285
CD	–500	–0.8222	0.001644	+285
AD	–2,000	–10.6856	0.005343	–1,225
		–10.64	0.007420	

$$\text{Flow correction} = \frac{-\Sigma(\text{HL})}{1.85\,\Sigma\,\text{HL/Q}} = \frac{-(-10.64)}{1.85 \times 0.007420\ \text{ft/gpm}}$$

= 775 gpm

THE CORRECT ANSWER IS: (D)

126. Addition of chlorine beyond the breakpoint increases the free chlorine residual.

THE CORRECT ANSWER IS: (B)

127. Free available chlorine = dose – demand = 1.8 – (0.45 × 1.8) = 0.99 mg/L

THE CORRECT ANSWER IS: (C)

128. The surface water treatment rule requires a 3.0 log inactivation of *Giardia*.

THE CORRECT ANSWER IS: (D)

129. Excessive phosphate in a lake contributes to aquatic plant growth and subsequent eutrophication.

THE CORRECT ANSWER IS: (A)

130. Phosphate is almost always the limiting nutrient (eliminates C & D). Nitrate, not calcium, is relevant (eliminates B).

THE CORRECT ANSWER IS: (A)

GO ON TO THE NEXT PAGE

131. Refer to LaGreg, M.D., Buckingham, P.L. and Evan, J.C., *Hazardous Waste Management*, McGraw-Hill, New York, New York, pg. 139, 1994.

THE CORRECT ANSWER IS: (A)

132. The predominant nitrogen species in anaerobic environments is NH_3. NO_3 is predominant in aerobic environments.

THE CORRECT ANSWER IS: (C)

133. Harvested algae may be used as a humic, N, and P amendment to soils. Option (C) is not correct because this is a method of disposal, not a use.

THE CORRECT ANSWER IS: (B)

134. $D_{MAX} = \frac{k}{k_2} L_0 e^{-kt_{max}}$

$$= \frac{0.23}{0.40} 10e^{-0.23(3\text{ days})}$$

$$= 2.88 \text{ mg/L}$$

Look up or calculate D.O. Saturation at 20°C = 9.02 mg/l
D.O. = $DO_{SAT} - D_{MAX}$ = 9.02 – 2.88 = 6.14 mg/l

THE CORRECT ANSWER IS: (C)

135. Using Figure 3.

Effective stack height = h + Δh (plume rise)

= (1.37 m)(12) + 8.5

= 16.5 + 8.5 = 25 m

X = 0.25 km or 250 m

THE CORRECT ANSWER IS: (B)

136. Emission rate in g/s

$$\frac{1{,}000\ \text{Btu}}{\text{ft}^3} \times \frac{379\ \text{ft}^3}{\text{lb mol CH}_4} \times \frac{\text{lb mol CH}_4}{16\ \text{lbm CH}_4} \times \frac{3{,}000\ \text{lbm CH}_4}{\text{hr}} \times \frac{0.15\ \text{lb NO}_x}{1\times10^6\ \text{Btu}} \times \frac{453.6\ \text{g NO}_x}{\text{lb NO}_x} \times \frac{1\ \text{hr}}{3{,}600\ \text{s}} = \frac{1.34\ \text{g}}{\text{s}}$$

Calculate effective stack height
$H = h + \Delta H = 16.44 + 8.5 = 24.94$ m

Determine stability class
Neutral Class D should be assumed for overcast conditions during day or night

From Figure 1 $\sigma_z = 90$ m
$\sigma_y = 350$ m

$$X = \frac{Q}{\pi\sigma_y\sigma_z\mu}\left[\exp\left[-\frac{1}{2}\left(\frac{y}{\sigma_y}\right)^2\right]\right]\left[\exp\left[-\frac{1}{2}\left(\frac{H}{\sigma_z}\right)^2\right]\right]$$

$$= \frac{1.343}{\pi(350)(90)(2.5)}\left[\exp\left[-\frac{1}{2}\left(\frac{0}{350}\right)^2\right]\right]\left[\exp\left[-\frac{1}{2}\left(\frac{24.94}{90}\right)^2\right]\right]$$

$$= 5.22\times10^{-6}\ \frac{\text{g}}{\text{m}^3} = 5.22\ \frac{\mu\text{g}}{\text{m}^3}$$

THE CORRECT ANSWER IS: (B)

137. The chemiluminescence method is the benchmark for measuring NO_x.

THE CORRECT ANSWER IS: (D)

138. There are no TLVs for concentrations of total culturable or countable organisms and particles because:

1) culturable organisms or countable spores do not compromise a single entity

2) human responses to bioaerosols range from innocuous effects to serious disease and depend on the specific agent and susceptibility of the person

3) measured concentrations are dependent on the method of sample collection and analysis

THE CORRECT ANSWER IS: (D)

139. Federal air quality standards address;

Particulates
Sulfur Dioxide
Nitrogen Oxides
Carbon Monoxide
Ozone
Lead

THE CORRECT ANSWER IS: (C)

140. Aerodynamic diameter = Da = $k\ d\ p^{1/2}$;

where
d = particle physical diameter
p = particle density

Therefore, $10 \times 1^{1/2} = 10$ and $5 \times 4^{1/2} = 10$

THE CORRECT ANSWER IS: (B)

141. Dispersion modeling along the centerline for downwind.
Ground level concentration can be expressed as:

$X(x,y) = [Q / (3.14\ (d_y d_z)\ u)]\ \exp\{-1/2(H/d_z)^2\}$

where
d_y and d_z = dispersion coefficients
Q = emission rate
u = mean wind speed
H = stack height

As the stack height increases the concentration decreases at the ground surface.

THE CORRECT ANSWER IS: (A)

142. NAAQS $NO_2 = 0.053\ ppm_v$ annual average

Max ambient receptor concentration $= \dfrac{220\,\mu g}{m^3}$

Receptor elevation $= 6{,}500$ ft

$P_{bar} = 23.4$ in Hg

$T = 70°F$

$$C_{mass} = \frac{P}{RT} C_{VOL}\, MW_p$$

$$= \frac{23.4 \text{ in. Hg}}{\frac{21.85 \text{ in. Hg}}{\text{lb mol°R}}(70+460)\text{°R}} \left(\frac{0.053}{1\times10^{6}}\right)\left(\frac{46\text{ lb}}{\text{lb mol}}\right) = 4.926\times10^{-9}\frac{\text{lb}}{\text{ft}^3}$$

$$C_{mass} = 4.926\times10^{-9}\frac{\text{lb}}{\text{ft}^3}\times\frac{4.536\times10^{8}\,\mu g}{\text{lb}}\times\frac{35.31\text{ft}^3}{\text{m}^3}$$

$$= 78.9\ \mu g/m^3$$

THE CORRECT ANSWER IS: (B)

143. $\dfrac{0.002}{760}\times 1{,}000{,}000 = 2.6$ ppm

THE CORRECT ANSWER IS: (A)

144. $d_{50} = 4$ microns

$\sigma_g = 2$ microns

$\log(d_{84.1}) = \log(d_{50}) + \log(\sigma_g)$

$= \log(4) + \log(2)$

$d_{84.1} = 10^{0.9031}$

$= 8.0$ microns

THE CORRECT ANSWER IS: (D)

145. Particulate emission rate = (emission factor)(coal flow rate)

= 139.2 lb/ton × 40 tons/hr = 5,568 lb/hr = 2.78 tons/hr

THE CORRECT ANSWER IS: (A)

146. $V = gD^2 \dfrac{\rho_{part} - \rho_{fluid}}{18\mu}$

$$= \frac{\left(9.81\ \text{m/s}^2\right)\left(5\times10^{-6}\right)^2\left(2{,}000-1.2\right)}{18\left(1.8\times10^{-5}\right)}$$

$= 1.51 \times 10^{-3}$ m/s

THE CORRECT ANSWER IS: (C)

147. $N_e = \dfrac{1}{H}\left(L_b + \dfrac{L_c}{2}\right)$

$$= \frac{1}{0.6}\left(1.5 + \frac{2.5}{2}\right)$$

$= 4.58$

Cooper & Alley, 2nd ed., pg. 132

THE CORRECT ANSWER IS: (B)

148. $N_2 = N_1\left(\dfrac{Q_2}{Q_1}\right)$

$$= \left(950\right)\left(\frac{1{,}000}{725}\right) = 1{,}310$$

THE CORRECT ANSWER IS: (D)

149. $\Delta h_r = \Delta h_{f,\ CO_2} + 2\Delta h_{f,\ H_2O} - \Delta h_{f,\ CH_4} - 2\Delta h_{f,\ O_2}$

$= (-394{,}088) + 2(-242{,}174) - (-74{,}980)$

$= 803{,}456$

THE CORRECT ANSWER IS: (C)

150. $0.30(45) + 0.10(50) + 0.60(90) = 72.50$

THE CORRECT ANSWER IS: (C)

SOLUTIONS FOR SAMPLE QUESTIONS IN THE AFTERNOON PORTION OF THE EXAMINATION IN ENVIRONMENTAL ENGINEERING

Answer Key for Afternoon Sample Questions

501	C	**526**	A
502	B	**527**	C
503	B	**528**	A
504	A	**529**	C
505	A	**530**	D
506	B	**531**	D
507	C	**532**	C
508	C	**533**	D
509	B	**534**	B
510	C	**535**	A
511	A	**536**	D
512	C	**537**	B
513	C	**538**	B
514	A	**539**	D
515	D	**540**	D
516	C	**541**	B
517	D	**542**	C
518	C	**543**	B
519	B	**544**	D
520	D	**545**	D
521	C	**546**	A
522	D	**547**	B
523	B ~~C~~	**548**	D
524	D	**549**	B
525	B	**550**	A

501. 6 ft × 0.5 ft × 1 fps = 3 cfs

3 cfs × 1.6 lb/ft^3 = 4.8 lb/sec

4.8 lb/sec × 6.25% = 0.3 lb H_2O/sec

The water to be evaporated is 0.3 lb/sec
The blanket material and adhesive combination appear at 4.5 lb/sec and must be heated to 400°F.

Water Evaporation Heat Requirement: 0.3 × (975 – 50) = 277.5 Btu/sec
Blanket Drying and Curing Heat Requirement: 4.5 × 2 × (400 – 75) = 2,925 Btu/sec

Together, expressed as Btu/hr this becomes: 3,202.5 × 3,600 = 11,529,000 Btu/hr

THE CORRECT ANSWER IS: (C)

502. From Hazardous Waste Management by Grega, et.al., pg. 774–5

Combustion Efficiency (CE) = (CO_2 @ 7%) / (CO @ 7% + CO_2 @ 7%)

Therefore, 100 × (13.9 / (21 – 12)) = 155 ppm CO corrected to 7% O_2

CE = 5,000 / (5,000 + 155) = 96.99%

THE CORRECT ANSWER IS: (B)

503. The standard flow rate can be calculated as follows:

$dq/dt = A \times V \times (P_s / P_{std}) \times (T_{std}\ °R / T_s\ °R) \times (1 - (\text{Moisture fraction})) \times 60$ sec/min

$$= \left[\frac{(92/12)^2}{4} \times \pi\right] ft^2 \times 28.84 \text{ fps} \times (23.56/29.92) \times (460 + 70) / (460 + 355) \times (1 - 0.1225) \times 60 \text{ sec/min}$$

= 35,895 Dry Standard Cubic Feet per Minute (DSCFM)

The emission rate of NO_x (lb/hr) can be calculated as:

= 117.27 ppmv × 1.194 × 10^{-7} lb NO_x / scf ppmv × dq/dt scf/min × 60 min/hr

= 30 lb NO_x /hr

503. (Continued)

The heat input rate for the boiler can be calculated as:

$= 149.9 \times 10^3$ scf/hr × 1,000 Btu/scf $= 149.9 \times 10^6$ Btu/hr

The emissions per 10^6 Btu then become:

$= 30$ lb/hr / 149.9×10^6 Btu/hr $= 0.2012$ lb NO_x / 10^6 Btu which exceeds the standard

THE CORRECT ANSWER IS: (B)

504. The correct answer is Best Available Control Technology. Source Pollution Control Guide is a publication, not a control technology. Lowest Achievable Emissions Rate required controls for criteria pollutants for modifications in non-attainment areas. Maximum Achievable Control Technology industry specific controls for hazardous air pollutants.

THE CORRECT ANSWER IS: (A)

505.

Newspaper:	6.4 × 7,200	=	46,100 Btu/100 lb
Corrugated Boards:	9.0 × 7,000	=	63,000
Ferrous:	5.0 × 300	=	1,500
Aluminum:	1.0 × 0	=	0
Glass:	6.5 × 60	=	390
			110,990 Btu/100 lb

$$\left(\frac{110{,}990 \text{ Btu}}{100 \text{ lb}}\right)(0.50)(0.75)(600 \text{ tpd})(2{,}000 \text{ lb/ton})$$

$= \quad 499.5 \times 10^6$ Btu/day

THE CORRECT ANSWER IS: (A)

506. (500 tpd)(0.40)(0.065)(30 \$/ton) + (500 tpd)(0.40)(0.01)(100 \$/ton)

\$390 + \$200 = \$590/day

THE CORRECT ANSWER IS: (B)

GO ON TO THE NEXT PAGE

507. (450 kWh/ton)(600 tpd)(365 days/year)(0.95)($0.04 / kWh)

$3,745,000/year gross revenue

THE CORRECT ANSWER IS: (C)

508. (1.5 in.)(10 acre)(43,560 ft^2/acre)(ft/12 in.) = 54,450 ft^3/day

= 0.63 cfs

THE CORRECT ANSWER IS: (C)

509. $V = \frac{1.486}{n} R^{2/3} s^{1/2}$

$$Q = VA = V\frac{\pi D^2}{4}$$

$$V = \frac{4Q}{\pi D^2} = \frac{(4)(0.45)}{3.14\, D^2} = \frac{0.57}{D^2}$$

$$\frac{0.57}{D^2} = \left(\frac{1.486}{n}\right)\left(\frac{\frac{\pi D^2}{4}}{\pi D}\right)^{2/3}(0.012)^{1/2}$$

$$\frac{0.57}{D^2} = \left(\frac{1.486}{n}\right)\left(\frac{D}{4}\right)^{2/3}(0.012)^{1/2}$$

D = 0.387 ft or 4.6 in.; diameter must be at least 6 in.

The answer is a 6-in.-diameter pipe.

THE CORRECT ANSWER IS: (B)

510. $C_t = C_o\ e^{-kt}$

$$\frac{C_t}{C_o} = e^{-kt}$$

$$\ln\frac{C_t}{C_o} = -kt$$

$$k = 0.002\ \text{day}^{-1}$$

$$C_o = 17\ \text{mg/kg}$$

$$C_t = 5\ \text{mg/kg}$$

$$t = \frac{\ln\left(\frac{C_t}{C_o}\right)}{-k}$$

$$= \frac{\ln\left(\frac{5}{17}\right)}{-0.002}$$

$$= \frac{\ln 0.294}{-0.002}$$

$$= \frac{-1.22}{-0.002}$$

$$= 610\ \text{days}$$

$$= 1.67\ \text{years}$$

THE CORRECT ANSWER IS: (C)

511. Although all are potential sources of leachate, the design of the liner system should preclude IV and V. Proper cell design should eliminate overland flow of water from one landfill cell to another.

THE CORRECT ANSWER IS: (A)

512. Municipal wastewater sludges must be dewatered prior to landfill disposal, achieving a minimum solids concentration of 15 to 20% (US EPA, 40 CFR 258.28, Liquid Restrictions).

THE CORRECT ANSWER IS: (C)

513. The minimum requirement for final cover on a municipal landfill is 24 in. (Salvato, pg. 665).

THE CORRECT ANSWER IS: (C)

514. $\frac{x}{500} = \frac{0.90}{0.475}$

$x = 947.4$ Btu/scf

let $M = \%$ methane

$P = \%$ propane

Therefore,

$$\begin{aligned} 2{,}350P + 947.4M &= 1{,}000 \\ P + M &= 1 \\ 2{,}350P + 947.4(1 - P) &= 1{,}000 \\ 2{,}350P + 947.4 - 947.4P &= 1{,}000 \\ 1{,}402.6P &= 52.6 \\ P &= 0.0375 \text{ or } 3.75\% \end{aligned}$$

THE CORRECT ANSWER IS: (A)

515. $$DRE_{chlorobenzene} = \left(1 - \frac{0.01}{153}\right) \times 100 = 99.993\%$$

$$DRE_{toluene} = \left(1 - \frac{0.037}{432}\right) \times 100 = 99.991\%$$

$$DRE_{xylene} = \left(1 - \frac{0.070}{435}\right) \times 100 = 99.984\%$$

THE CORRECT ANSWER IS: (D)

516. W_{CB} = flow of chlorobenzene (kg/hr)

M_{CB} = molar flow rate of chlorobenzene (moles/hr)

MW_{CB} = molecular weight of chlorobenzene (g/mole)

$$M_{CB} = \frac{W_{CB}}{MW_{CB}} = \frac{(153 \text{ kg/hr})(1{,}000 \text{ g/kg})}{MW_{CB}}$$

$MW_{CB} = (6 \times 12) + (5 \times 1) + (35.5)$

$= 112.5$ g/mole

$$M_{CB} = \frac{153{,}000}{112.5} = 1{,}360 \text{ moles/hr}$$

Note: Each mole of chlorobenzene contains one atom of chlorine, therefore:

$M_{HCl} = M_{CB} = 1{,}360$ moles/hr

W_{HCl} = flow of HCl

MW_{HCl} = molecular weight of HCl

$W_{HCl} = MW_{HCl}$ (moles/hr)

$W_{HCl} = (35.5 + 1)(1{,}360)$

$W_{HCl} = 49{,}640$ g/hr or 49.64 kg/hr

THE CORRECT ANSWER IS: (C)

517. $$M_{HCl} = M_{CB} = \frac{(W_{CD} \text{ kg/hr})(1{,}000 \text{ g/kg})}{MW_{CB}}$$

$W_{HCl} = MW_{HCl} \times M_{CB}$

1% uncontrolled emission = $(0.01)(W_{HCl}) = 1.8$ kg/hr

Therefore, $$1.8 = (0.01)(MW_{HCl})\left[\frac{W_{CD}}{MW_{CB}}\right]$$

517. (Continued)

$$1.8 = (0.01)(36.5)\left[\frac{W_{CD}}{112.5}\right] = (g/mole)\left[\frac{kg/hr}{g/mole}\right]$$

$$W_{CD} = \frac{(1.8)(112.5)}{(0.01)(36.5)} = 554.8 \text{ kg/hr}$$

THE CORRECT ANSWER IS: (D)

518. Although flexible membrane liners can be used on very steep slopes, the manufacturers recommend a slope of 3:1 (H:V) for general application.

THE CORRECT ANSWER IS: (C)

519. The "witness zone" is that section of the landfill that ascertains whether the upper or primary flexible membrane liner is leaking. Thus the section between the lower and upper flexible membrane liners is the "witness zone."

THE CORRECT ANSWER IS: (B)

520. The compacted low-permeability soil, usually clay, should be tested to have an in-place conductivity of no less than 1×10^{-7} cm/s.

THE CORRECT ANSWER IS: (D)

521. (A) and (B) are not acceptable solutions under RCRA; therefore, "recycling" internally is the only option.

THE CORRECT ANSWER IS: (C)

522. $$\frac{(700 \text{ gal/month})(1.1)(8.347 \text{ lb/gal})}{2{,}200 \text{ lb/metric ton}} = 2.921 \text{ metric tons/month}$$

This is a large generator. The generator must obtain a permit to store waste over 90 days. Thus, the manufacturer is a large generator, not in compliance.

THE CORRECT ANSWER IS: (D)

523. 4,200 gal of chlorobenzene

$(4{,}200)(8.34)(1.1)(1/2.2)(0.04) = 700.6 \text{ kg } C_6H_5Cl$

$$99.99 = 100\left(1 - \frac{W_{out}}{W_{in}}\right) = 100\left(1 - \frac{W_{out}}{700.6}\right)$$

$$0.9999 = 1 - \frac{W_{out}}{700.6}$$

$w_{out} = (1 - 0.9999)(700.6) = 0.07006$

THE CORRECT ANSWER IS: (C) B

524. i = 7%
n = 5 years
P = $245,000

pg 54-16
App 54.B (A/P, 7%, 5)

From Interest Tables, A = 0.2439

Annualized Cost = 0.2439 (245,000) + 9,000
= 59,756 + 9,000
= $68,756

THE CORRECT ANSWER IS: (D)

525. High molecular weight compounds with hydrophobic non-polar organic matter are highly partitioned (attached) to other organic materials such as humic soil matter. This process is called sorption and is based on relative affinity, which is primarily a molecular phenomenon and is a function of multiple chemical, physical, and electrostatic mechanisms.

THE CORRECT ANSWER IS: (B)

526. Microorganisms are readily transported through desiccation (cracks and fissures) and rainfall. The amount of microorganisms applied is totally irrelevant as is the pH of the soil. Viruses, although smaller than other microorganisms, will not be transported without an aqueous environment.

THE CORRECT ANSWER IS: (A)

527. Terrain conductivity and resistivity are surface techniques, not downhole techniques. Ground penetrating radar is used for detection of underground objects, not soil continuity.

THE CORRECT ANSWER IS: (C)

528. Darcy velocity calculation:

$$V = K(dh/dr)$$

$$dh/dr = (0.30 + 1.08) / 1.08 = 1.28$$

$$V = (1 \times 10^{-7}) \times (1.28) = 1.28 \times 10^{-7} \text{ cm/sec}$$

Seepage velocity calculation:

$$V' = ((K(dh/dr)) / n$$

$$V' = (1.28 \times 10^{-7}) / 0.55 = 2.33 \times 10^{-7} \text{ cm/sec}$$

Travel time calculation:

$$t = \text{distance/velocity}$$

$$t = (1.08) \times (100) / (2.33 \times 10^{-7})$$

$$t = 4.64 \times 10^{8} \text{ seconds or 14.7 years}$$

THE CORRECT ANSWER IS: (A)

529. Movement of relatively non-soluble contaminants in groundwater is primarily accomplished by natural colloidal motion through the groundwater table. These contaminants attach themselves to small soil particles that are physically transported by groundwater travel.

THE CORRECT ANSWER IS: (C)

530. By definition, the correct answer is Option D.

THE CORRECT ANSWER IS: (D)

531. $$\text{HI} = \frac{\text{Actual MTBE intake}}{\text{Oral } R_f D}$$

$$\text{HI} = \frac{0.00275 \text{ mg}/(\text{kg} \cdot \text{day})}{0.005 \text{ mg}/(\text{kg} \cdot \text{day})} = 0.55$$

THE CORRECT ANSWER IS: (D)

532. $$\text{DWEL} = \frac{(R_f D_{oral})(\text{ABW})}{(\text{DWI})(\text{AB})(\text{FOE})}$$

DWEL = Drinking water equivalent level, mg/L

$R_f D_{oral}$ = Oral $R_f D$ for MTBE, mg/(kg•day)

ABW = Average body weight, kg

DWI = Daily water intake, L

AB = Absorbed dose (1.0)

$$\text{FOE} = \frac{365 \text{ days}}{365 \text{ days}} = 1$$

$$\text{DWEL} = \frac{[0.005 \text{ mg}/(\text{kg} \cdot \text{day})](15 \text{ kg})}{(1 \text{ L}/\text{day})(1)(1)} = 0.075 \text{ mg}/\text{L}$$

THE CORRECT ANSWER IS: (C)

533. Removal with soil vapor extraction wells is the only effective technology for removing organic contaminants from soil systems. Pump and treat methods refer to contaminated groundwater. Soil washing with low pH solutions is effective only for the subsurface soils in contact with the solution and may also introduce additional contaminants. Injection of bacteria cultures is only potentially effective for relatively low-concentration ranges of contaminants.

THE CORRECT ANSWER IS: (D)

534. Evaluate the four options:

Option (A) — alfalfa and clover are legumes and should never receive nitrogen under nitrogen-limited conditions.

Option (B) — lime stabilization and no incorporation
$(0.30)(0.25) = 0.075$

Option (C) — anaerobic digestion and incorporation within 1–7 days
$(0.20)(0.70) = 0.14$

Option (D) — aerobic digestion and injection
$(0.30)(0.95) = 0.27$

The least availability is with lime stabilization and no incorporation.

THE CORRECT ANSWER IS: (B)

535.

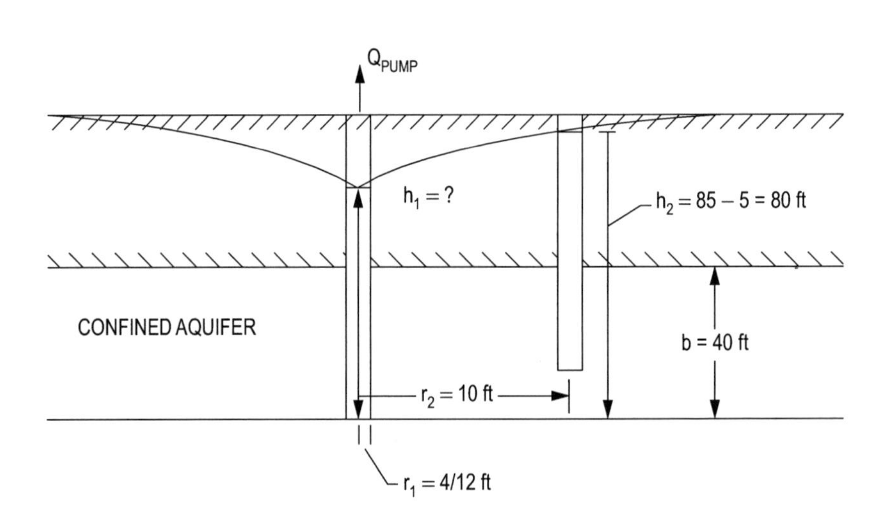

Steady state flow in a confined aquifer is determined by some form of the Thiem Equation.

From Fetter, 1994, pg. 217.

$$h_2 - h_1 = \frac{Q}{2\pi T} \ln(r_2/r_1)$$

h_2, h_1 = hydraulic head, ft

Q = pumping rate, gpd
= 35 g/min × 60 min/hr × 24 hr/day = 50,400 gpd

r_1, r_2 = radius distance from center of well, ft

535. (Continued)

T = transmissivity, gal/(day-ft)
= hydraulic conductivity × thickness
= 175 gal/(day-ft^2) × 40 ft
= 7,000 gal/(day-ft)

$$80 - h_1 = \left(\frac{50{,}400 \text{ gpd}}{(2\pi)(7{,}000 \text{ gal}/(\text{day-ft}))} \right) \ln\left[\frac{10}{0.33}\right]$$

$$80 - h_1 = (1.146)(3.40)$$

$$h_1 = 76.1 \text{ ft}$$

However, drawdown was asked for,

Drawdown = 85.0 – 76.1 = 8.9 ft

THE CORRECT ANSWER IS: (A)

536. Air Space Saved = ((200 ft)(1,000 ft)(15 ft + 20 ft)(15%)) / 27 ft^3/yd^3 = 38,889 yd^3

Revenue from Added Air Space = (38,889 yd^3)(\$25.00/yd^3) = \$972,222

Net Revenue Increase = \$972,222 – \$250,000 = \$722,222

Savings Achievable with Alternate Cover = \$722,222 + \$100,000 = \$822,222

THE CORRECT ANSWER IS: (D)

537. When the mixed waste no longer retains the characteristic that made it hazardous and it has no other hazardous waste characteristic, the combined waste is no longer classified as a hazardous waste.

Reference: Travis Wagner, *Hazardous Waste Regulations*, 2nd ed., pp 37–38

THE CORRECT ANSWER IS: (B)

GO ON TO THE NEXT PAGE

538. Incinerator Efficiency required by RCRA is 99.99%

Chlorobenzene (CB) at inlet = 1,200 L/10^6 liters (ppmv) × (2 × 10^6 L/hr) = 2,400 L/hr CB

At 99.99% removal: 0.0001 × 2,400 = 0.24 L/hr CB in outlet gases at inlet conditions

PV = (gms CB/MW CB) × R × T

780 × 0.24 = (gms CB/112.5) × 62.361 × 421.9

where at inlet conditions:

P = 780 mm Hg
Mole weight = 112.5 gms
Gas law constant, R = 62.361 (from Perry's Handbook, 4th ed., pg. 1–25)
Temperature = 421.9 K

Therefore, gms CB = 780 × 0.24 × 112.5 / (62.361 × 421.9) = 0.8 gms

THE CORRECT ANSWER IS: (B)

539. More MTBE would NOT be removed because the equilibrium of adsorption would drive MTBE to the vapor phase at higher temperatures.

THE CORRECT ANSWER IS: (D)

540. Necessary Spent Bed Volume $= \dfrac{75{,}000 \text{ lb}}{259 \times 10^{-3} \text{lb/cm}^3} = 28.958 \times 10^6 \text{ cm}^3$

Using the same bed thickness as the sample bed we get:

$$\text{Bed Surface Area} = \frac{28.958 \times 10^6 \text{ cm}^3}{2.15 \text{ cm}} = 13.469 \times 10^6 \text{ cm}^2$$

$$= \frac{13.469 \times 10^6 \text{ cm}^2}{(2.54 \text{ cm/in.})(2.54 \text{ cm/in.})(12 \text{ in./ft})(12 \text{ in./ft})} = 14{,}497.5 \text{ ft}^2$$

THE CORRECT ANSWER IS: (D)

GO ON TO THE NEXT PAGE

541. The designation "confined space" is independent of hazard. "Permit required confined space" is a confined space with potential for IDLH.

THE CORRECT ANSWER IS: (B)

542. SCBA operated in the pressure demand mode is the only respirator approved for protection against unknowns. Sampling should account for pockets of high concentration.

THE CORRECT ANSWER IS: (C)

543. 19.5% or less is commonly defined as an oxygen-deficient atmosphere.

THE CORRECT ANSWER IS: (B)

544. The worker has worked a 12-hr shift, so the total exposure is:

$(15\ \text{mg/m}^3)(4\ \text{hr}) + (0)(1) + (7)(3) + (9)(4) = 117\ \text{mg/(m}^3\cdot\text{hr)}$,

and the TWA (8 hr) $= \dfrac{117\ \text{mg}/\left(\text{m}^3\cdot\text{hr}\right)}{8\ \text{hr}} = 14.6\ \text{mg}/\text{m}^3$

Source: ACGIH Threshold Limit Values

THE CORRECT ANSWER IS: (D)

545. HVL is the thickness of a material required to reduce radiation intensity by a factor of two, and shield can be expressed mathematically in terms of HVLs as:

Fraction of original level = $1/2^n$

To reduce the radiation level to 2.5 mrem/hr is a four-fold reduction, hence

$1/4 = 1/2^n$

$2^n = 4$

$n = 2$ HVL or $2 \times 0.33 = 0.66$ cm

Source: *Radiological Health Handbook*

THE CORRECT ANSWER IS: (D)

GO ON TO THE NEXT PAGE

546. $C = C_o e^{-kt}$

$$t = -\frac{\ln(C/C_o)}{k}$$

$$= -\frac{\ln(0.5)}{0.0175}$$

$$= 39.6 \text{ days}$$

THE CORRECT ANSWER IS: (A)

547. $\Sigma = 10^{100/10}(0.091) + 10^{60/10}(0.91)$

$$= 9.1\times10^8 + 9.1\times10^5$$

$$= 9.109\times10^8$$

$$L_{eq} = 10\log(9.109\times10^8)$$

$$= 89.6 \text{ dBA}$$

THE CORRECT ANSWER IS: (B)

548. A mass balance for the hydrocarbon gives:

$$V\frac{dc}{dt} = C_aQ - QC$$

$$= -Q(C - C_a)$$

where

V = volume of enclosed space

$\frac{dc}{dt} = \frac{\text{change in concentration}}{\text{change in time}}$

C_a = outside air concentration

Q = volumetric flow rate

Integrating gives:

$$\frac{5}{100{,}000} = \exp\frac{-t\times28.32}{150}$$

$$= 52.44 \text{ min}$$

Using a safety factor of 4 gives a time of 210 min.

THE CORRECT ANSWER IS: (D)

549. First determine the amount of contaminant:

$$32 \text{ oz} \times \frac{\text{lb}}{16 \text{ oz}} \times 0.87 \times \frac{453{,}600 \text{ mg}}{\text{lb}} = 789{,}264 \text{ mg}$$

Next determine the volume of the room:

$$20 \text{ ft} \times 30 \text{ ft} \times 10 \text{ ft} \times 0.02832 \text{ m}^3/\text{ft}^3 = 170 \text{ m}^3$$

Next determine the theoretical max concentration in the room:

$$\frac{789{,}264 \text{ mg}}{170 \text{ m}^3} = 4{,}643 \text{ mg/m}^3$$

Last check to see if saturation limit is exceeded:

$$\frac{21}{760} \times 1{,}000{,}000 \times \frac{92}{24.45} = 103{,}971 \text{ mg/m}^3$$

THE CORRECT ANSWER IS: (B)

550. Both nitrile and polyvinyl alcohol material will give acceptable protection for MTBE, but polyvinyl alcohol will dissolve in water. Since the spill is in a river, polyvinyl alcohol material is not acceptable. Cotton and neoprene are not acceptable due to insufficient protection time.

THE CORRECT ANSWER IS: (A)

Appendix A

Sample of exam covers and instructions

Serial No. SAMPLE

Name: ____
Last First Middle Initial

Principles and Practice of Engineering

ENVIRONMENTAL ENGINEERING

MORNING SESSION—0404

READ THE FOLLOWING INSTRUCTIONS CAREFULLY.

- If you do not comply with these instructions, your examination score may be **INVALIDATED.**
- Darken **ALL** answers on the answer sheet enclosed within this examination book. Only the answer sheet will be scored. The answer sheet is the only record of your answers.
- **COMPLETELY DARKEN** the circles corresponding to the answers you choose. Use **ONLY** the pencil provided.
- Do all scratch work only in the blank spaces in this examination book. You may **NOT** write on loose paper or in reference books. **NO CREDIT** will be given for any work written in the examination book. You will **NOT** be given extra time to transfer answers to the answer sheet.
- This is an open-book examination. All reference materials must be bound as a book or in a three-ring binder. Loose papers, writing tablets, unbound notes or tables, and devices that may compromise the security of the examination are **NOT PERMITTED** in the examination room. Prohibited items will be collected by a proctor.
- Do not share or exchange reference materials with other examinees. You may use **ONLY** your reference materials.
- Devices or materials that might compromise the security of the examination or examination process are **NOT PERMITTED**. Calculators with communication or text-editing capabilities are prohibited, as are communication devices such as pagers and cellular phones. Calculating and computing devices having a QWERTY keypad arrangement similar to a typewriter are not permitted. These devices include but are not limited to palmtop, laptop, and handheld computers, calculators, databanks, data collectors, and organizers. Prohibited items will be collected by a proctor.
- Copying examination questions for future reference, recording examination questions into an electronic device, copying answers from other examinees, or cheating of any kind is **NOT PERMITTED** and will result in an invalidation of your examination score.
- The morning examination is a maximum of four hours in length. Work all **50** questions according to the instructions. All questions are equally weighted. Points are not subtracted for incorrect responses.
- Only one answer is permitted for each question; no credit is given for multiple answers. If you change an answer, be sure to completely erase the previous mark. Incomplete erasures may be read as intended answers.
- At the conclusion of the examination, you are responsible for returning the numbered examination book that was assigned to you.

Instructions for Completing Answer Sheet

1. When the proctor prompts you, slide the answer sheet out from under the front cover of this examination book. **DO NOT** break the seal on the examination book. **DO NOT** open the examination book.

2. Boxes 1 and 2 request that you print your name, exam date, and exam location. Print legibly and neatly so that the answer sheet can be identified.

3. Box 3 contains a very important agreement between you and NCEES and your local licensure board. If you decide not to sign the agreement, raise your hand and a proctor will collect your examination materials uncompleted. By signing the agreement, you are legally bound to abide by the terms of the agreement. If you do not abide by the terms, your exam score may be invalidated, and/or you may be barred from retaking the exam. These terms include affirmation that the answers you provide are solely of your knowledge and hand; you will not copy any information onto material to be taken from the exam room; you will not reveal in whole or in part any exam questions, answers, problems, or solutions to anyone during or after the exam, whether orally, in writing, or in any Internet "chat rooms" or otherwise. Once you have read, understood, and agreed to the terms of the agreement, sign your name in box 3.

4. Complete the information requested in boxes 4 and 5 of the answer sheet. If the day of your birth is one digit, add a zero in front of the digit so that all boxes are filled. This information is very important for identification purposes.

5. Box 6 requests your examinee identification (I.D.) number. Incorrectly entering your I.D. number may delay your score. If necessary, add leading zeros to your I.D. number so that every box is filled. Do not include letters in your I.D. number. For example, if your I.D. number is T1580, enter the number like this:

0	0	0	0	0	1	5	8	0

   Complete the information requested in boxes 7–10. This information is very important for statistical purposes. (Disregard box 11 at this time.)

6. Turn over the answer sheet. In box 12, darken the appropriate circle.

7. In box 13, duplicate the examinee identification number you provided in box 6. Be certain that the examinee identification number you provide is correct and dark enough to photocopy.

8. Box 14 requests the examination book serial number. This number is found on the front top of this examination book.

9. In box 15, **COMPLETELY DARKEN** the circle corresponding to the examination that you choose to take. It is *very* important that the examination you work is the examination indicated on the answer sheet.

10. At the bottom of the answer sheet, locate the first four answer boxes numbered 101–150. These are the only boxes you will use to indicate your answers.

11. You will not receive additional time to transfer answers to the answer sheet. When the proctor says the exam has ended, you must put down your pencil and stop writing.

12. If you have a concern regarding the validity of an examination question, request an Examinee Comment Form from a proctor as you exit the room after you have completed the examination.

DO NOT OPEN THE EXAMINATION BOOK

UNTIL INSTRUCTED TO DO SO BY THE PROCTOR.

Serial No. SAMPLE

Name: ____
Last First Middle Initial

Principles and Practice of Engineering

ENVIRONMENTAL ENGINEERING

AFTERNOON SESSION—0404

READ THE FOLLOWING INSTRUCTIONS CAREFULLY.

- If you do not comply with these instructions, your examination score may be **INVALIDATED.**
- Darken **ALL** answers on the answer sheet enclosed within this examination book. Only the answer sheet will be scored. The answer sheet is the only record of your answers.
- **COMPLETELY DARKEN** the circles corresponding to the answers you choose. Use **ONLY** the pencil provided.
- Do all scratch work only in the blank spaces in this examination book. You may **NOT** write on loose paper or in reference books. **NO CREDIT** will be given for any work written in the examination book. You will **NOT** be given extra time to transfer answers to the answer sheet.
- This is an open-book examination. All reference materials must be bound as a book or in a three-ring binder. Loose papers, writing tablets, unbound notes or tables, and devices that may compromise the security of the examination are **NOT PERMITTED** in the examination room. Prohibited items will be collected by a proctor.
- Do not share or exchange reference materials with other examinees. You may use **ONLY** your reference materials.
- Devices or materials that might compromise the security of the examination or examination process are **NOT PERMITTED**. Calculators with communication or text-editing capabilities are prohibited, as are communication devices such as pagers and cellular phones. Calculating and computing devices having a QWERTY keypad arrangement similar to a typewriter are not permitted. These devices include but are not limited to palmtop, laptop, and handheld computers, calculators, databanks, data collectors, and organizers. Prohibited items will be collected by a proctor.
- Copying examination questions for future reference, recording examination questions into an electronic device, copying answers from other examinees, or cheating of any kind is **NOT PERMITTED** and will result in an invalidation of your examination score.
- The afternoon examination is a maximum of four hours in length. Work all **50** questions according to the instructions. All questions are equally weighted. Points are not subtracted for incorrect responses.
- Only one answer is permitted for each question; no credit is given for multiple answers. If you change an answer, be sure to completely erase the previous mark. Incomplete erasures may be read as intended answers.
- At the conclusion of the examination, you are responsible for returning the numbered examination book that was assigned to you.

INSTRUCTIONS FOR COMPLETING ANSWER SHEET

13. When the proctor prompts you, slide the answer sheet out from under the front cover of this examination book. **DO NOT** break the seal on the examination book. **DO NOT** open the examination book.

14. Boxes 1 and 2 request that you print your name, exam date, and exam location. Print legibly and neatly so that the answer sheet can be identified.

15. Box 3 contains a very important agreement between you and NCEES and your local licensure board. If you decide not to sign the agreement, raise your hand and a proctor will collect your examination materials uncompleted. By signing the agreement, you are legally bound to abide by the terms of the agreement. If you do not abide by the terms, your exam score may be invalidated, and/or you may be barred from retaking the exam. These terms include affirmation that the answers you provide are solely of your knowledge and hand; you will not copy any information onto material to be taken from the exam room; you will not reveal in whole or in part any exam questions, answers, problems, or solutions to anyone during or after the exam, whether orally, in writing, or in any Internet "chat rooms" or otherwise. Once you have read, understood, and agreed to the terms of the agreement, sign your name in box 3.

16. Complete the information requested in boxes 4 and 5 of the answer sheet. If the day of your birth is one digit, add a zero in front of the digit so that all boxes are filled. This information is very important for identification purposes.

17. Box 6 requests your examinee identification (I.D.) number. Incorrectly entering your I.D. number may delay your score. If necessary, add leading zeros to your I.D. number so that every box is filled. Do not include letters in your I.D. number. For example, if your I.D. number is T1580, enter the number like this:

0	0	0	0	0	1	5	8	0

18. In box 7, darken the appropriate circle.

19. Turn over the answer sheet. In box 8, duplicate the examinee identification number you provided in box 6. Be certain that the examinee identification number you provide is correct and dark enough to photocopy.

20. Box 9 requests the examination book serial number. This number is found on the front top of this examination book.

21. In box 10, **COMPLETELY DARKEN** the circle corresponding to the examination that you choose to take. It is *very* important that the examination you work is the examination indicated on the answer sheet.

22. Locate the first four answer boxes numbered 501–550. These are the only boxes you will use to indicate your answers.

23. You will not receive additional time to transfer answers to the answer sheet. When the proctor says the exam has ended, you must put down your pencil and stop writing.

24. If you have a concern regarding the validity of an examination question, request an Examinee Comment Form from a proctor as you exit the room after you have completed the examination.

DO NOT OPEN THE EXAMINATION BOOK UNTIL INSTRUCTED TO DO SO BY THE PROCTOR.